Frontiers in Mathematics

Advisory Editors

William Y. C. Chen, Nankai University, Tianjin, China

Laurent Saloff-Coste, Cornell University, Ithaca, NY, USA

Igor Shparlinski, The University of New South Wales, Sydney, NSW, Australia

Wolfgang Sprößig, TU Bergakademie Freiberg, Freiberg, Germany

This series is designed to be a repository for up-to-date research results which have been prepared for a wider audience. Graduates and postgraduates as well as scientists will benefit from the latest developments at the research frontiers in mathematics and at the "frontiers" between mathematics and other fields like computer science, physics, biology, economics, finance, etc. All volumes are online available at SpringerLink.

Elisa Affili • Serena Dipierro • Luca Rossi •
Enrico Valdinoci

A New Lotka-Volterra Model of Competition With Strategic Aggression

Civil Wars When Strategy Comes Into Play

Elisa Affili
Laboratoire de Mathématiques Raphaël Salem
Université de Rouen Normandie
Rouen, France

Serena Dipierro
Department of Mathematics and Statistics
University of Western Australia
Crawley, WA, Australia

Luca Rossi
Dipartimento di Matematica
Sapienza Università di Roma
Roma, Italy

Enrico Valdinoci
Department of Mathematics and Statistics
University of Western Australia
Crawley, WA, Australia

ISSN 1660-8046   ISSN 1660-8054   (electronic)
Frontiers in Mathematics
ISBN 978-3-031-67209-5   ISBN 978-3-031-67210-1   (eBook)
https://doi.org/10.1007/978-3-031-67210-1

Mathematics Subject Classification: 92D25, 37N25, 92B05, 34A26, 93B03

© The Editor(s) (if applicable) and The Author(s), under exclusive license to Springer Nature Switzerland AG 2024

This work is subject to copyright. All rights are solely and exclusively licensed by the Publisher, whether the whole or part of the material is concerned, specifically the rights of translation, reprinting, reuse of illustrations, recitation, broadcasting, reproduction on microfilms or in any other physical way, and transmission or information storage and retrieval, electronic adaptation, computer software, or by similar or dissimilar methodology now known or hereafter developed.
The use of general descriptive names, registered names, trademarks, service marks, etc. in this publication does not imply, even in the absence of a specific statement, that such names are exempt from the relevant protective laws and regulations and therefore free for general use.
The publisher, the authors and the editors are safe to assume that the advice and information in this book are believed to be true and accurate at the date of publication. Neither the publisher nor the authors or the editors give a warranty, expressed or implied, with respect to the material contained herein or for any errors or omissions that may have been made. The publisher remains neutral with regard to jurisdictional claims in published maps and institutional affiliations.

This book is published under the imprint Birkhäuser, www.birkhauser-science.com by the registered company Springer Nature Switzerland AG
The registered company address is: Gewerbestrasse 11, 6330 Cham, Switzerland

If disposing of this product, please recycle the paper.

Preface

In this monograph, we introduce a new model in population dynamics that describes two species, or communities, sharing the same environmental resources in a situation of open hostility.

Though the main methodology fits into the broad realm of mathematical biology, relying on methods from dynamical systems, ordinary differential equations, optimization, and optimal control, the model proposed and the results presented here are completely new.

The model is deduced from basic principles, accounting for competition and hostility between two species sharing the same environment. The fact of sharing this environment constitutes one of the salient features of the model in the description of a situation that, for human populations, is typical of civil wars. The model is also adapted to describe economic situations where two companies compete for the same market.

Our key assumption is that one of the two populations deliberately seeks for hostility through "targeted attacks." Hence the interaction is described not in terms of random encounters but rather via the strategic decisions of one population that can attack the other according to different levels of aggressiveness.

This leads to a non-variational model for the two populations in conflict, taking into account structural parameters such as the relative fit of the two populations with respect to the available resources and the effectiveness of the attack strikes of the aggressive population. One of the features that distinguishes this model from usual competitive systems is that it allows one of the population to go extinct *in finite time*.

The analysis that we perform focuses on the dynamical properties of the system, by detecting and describing all possible equilibria and their basins of attraction. Moreover, we analyze the strategies that may lead to the victory of the aggressive population, i.e., the choice of the aggressiveness parameter, in dependence of the structural constants of the system and possibly varying in time in order to make the attacks effective, which take to the extinction in finite time of the defensive population.

From the technical point of view, analyzing the case of two populations allows us to exploit techniques typical of two-dimensional dynamical systems, in which trajectories tend to separate different regions and in which asymptotic behaviors are regulated by the Poincaré-Bendixson Theorem (in particular, the system does not exhibit strange attractors). The analysis of the linear stability of the system, combined with a careful detection of stable and unstable orbits, also allows us to fully characterize the behavior of the equilibria, in dependence of the parameters involved in the description of the model (such as the fitness of the populations to the environment, the level of aggressiveness of the hostile population, and the effectiveness and cost of the attacks).

A number of bespoke analytical arguments are utilized to detect and identify the set of points for which finite time extinction is possible. In particular, different constraints on the aggressiveness parameter lead to different sets of initial points for which the victory is possible, highlighting that bang-bang strategies are necessary and sufficient to perform the task. We also analyze time minimizing strategies, for which variational tools such as the Pontryagin's Maximum Principle play a role in the optimality conditions.

Besides its mathematical interest, we think that the subject of this monograph is also topical and of great impact. Indeed, during the last century, the average duration of civil wars has significantly grown: since the end of World War II, this average duration has risen from about one-and-a-half-year to over 4 years (see [42]). This increased average length of civil wars has also resulted in an increased number of wars ongoing at any one time, thus contributing to a rise of tensions and potentially conflict situations, which may also contribute to the surge of new civil wars in a spiral of concurrent effects (for instance, about twenty contemporary civil wars took place close to the end of the Cold War).

Given the localized structure of these conflicts, civil wars often entail a large numbers of casualties, also among civilians (it is estimated that civil wars have caused the deaths of over 25 million people since 1945).

The intensity of the conflict and the high number of collateral damages typically cause a severe consumption of significant resources, a rise of speculative financial operations, and a radical pauperization of the territory.

Causes for civil wars are varifold, including ethnic and religious fractionalizations, poverty and social inequalities, governance and political issues. Classical analysis of civil wars focused on "greed versus grievance" as the two baseline arguments as causes of civil wars, where "greed" collects economic motivations making the best interests for individuals to join a rebellion and "grievance" is shorthand for all issues of ethnic, religious, and social tensions that contribute to the development of a conflict (see [20]).

Various causes to start and prolong civil wars have also been detected (see [57]), such as the possibility for elite groups to control economic resources and power positions, or to obtain private profit by mobilizing violent riots.

Among the myriad of factors that may stimulate or consolidate civil wars, we also mention the lack of accountability by political leaders (see [97]) and the size of a country's population (see [21]).

A strategical configuration of the territory can also be a reinforcing factor for a civil war: for instance, high levels of population dispersion and the presence of mountainous terrain can favor riots, making the population harder to control (see [21]).

Humans are not the only population that exhibit organized aggressive behavior. This tendency is common to several primate groups but also to evolutionarily very different species. In his work devoted to the study of ants, Wilson states: "The foreign policy aim of ants can be summed up as follows: restless aggression, territorial conquest, and genocidal annihilation of neighboring colonies whenever possible" (see [47]). Our model shows that this effect is not negligible in population dynamics models and opens the way for new studies in mathematics for ecology.

Besides the specific application in the study of aggressiveness phenomena in population dynamics, we stress that the model that we present here is also well-suited to describe confrontation between rival players or agents in various different contexts, such as strategic games or marketing models. For instance, starting from the Bass diffusion model [8], which describes the number of buyers or adopters of a product, and considering two products competing for the same market, leads to a Lotka–Volterra competitive system, see [15,59]. Our model then describes the situation where the two products are fabricated by distinct companies, and one of the two companies resorts to aggressive policies (such as misleading advertising, or releasing computer viruses) to set the rival product out of the market.

Rouen, France
Crawley, WA, Australia
Roma, Italy
Crawley, WA, Australia

Elisa Affili
Serena Dipierro
Luca Rossi
Enrico Valdinoci

Acknowledgments

It is a pleasure to thank Emmanuel Trélat for very interesting discussions.

Elisa Affili and Luca Rossi have been supported by the Italian INDAM-GNAMPA.

Serena Dipierro has been supported by the Australian Research Council DECRA DE180100957 *PDEs, free boundaries and applications*.

Luca Rossi has been supported by the French project ANR-23-CE40-0023-01 "REACH" *Réaction-diffusion: nouveaux défis*.

Enrico Valdinoci has been supported by the Australian Laureate Fellowship FL190100081 *Minimal surfaces, free boundaries and partial differential equations*.

Contents

1 **Introduction** 1
   1.1   Themes and Aims of This Work.. 2
   1.2   Disclaimer 3
   1.3   Organization of This Monograph 4

2 **Description of the Model** 5
   2.1   Motivations 6
   2.2   The Notion of Aggressiveness... 8
   2.3   Derivation of the Model 11
   2.4   Interpretation of the Model in an Economic Key 12

3 **Statement of the Main Results** 15
   3.1   Basic Results on the Dynamics... 15
   3.2   Constant Strategies 17
   3.3   Winning Strategies 20
   3.4   Time Minimizing Strategy 23
   3.5   Discussion of the Results 25

4 **Toolbox** 27

5 **Basins of Attractions** 39
   5.1   Characterization of $\mathcal{M}$ When $ac \neq 1$ 41
   5.2   Characterization of $\mathcal{M}$ When $ac = 1$ 49
   5.3   Study of the Dynamics... 55

6 **Parameters Dependence** 59
   6.1   Dependence on the Parameter c 59
   6.2   Dependence on the Parameter ρ 63
   6.3   Dependence on the Parameter a 69

7 Strategies of the First Population ... 85
7.1 Winning Nonconstant Strategies ... 85
7.2 Winning Strategies ... 91
7.3 The Role of the Constant Strategies ... 98
7.4 The Role of Heaviside Functions ... 126
7.5 Pointwise Constraints ... 127
7.6 Minimization of the War Duration ... 141

Bibliographical Notes ... 145

References ... 149

Introduction 1

Mathematical biology is a traditional field of investigation that bridges together different branches of pure and applied mathematics in strong connection with several disciplines in biology, such as ethology, behavioral ecology, cytology, evolutionary biology, cancer modeling, neuroscience, etc. (see [71, 72]).

The birth of mathematical biology dates back to the thirteenth century, when the famous Fibonacci sequence was introduced to calculate the growth of rabbit populations. The tradition of mathematical biology consolidated in the eighteenth and nineteenth centuries, also due to the works of Thomas Malthus, who described the population growth in terms of exponential functions, and Pierre François Verhulst, who introduced the mathematical notion of competition for resources and formulated the logistic equation (see, e.g., [24] for an overview of the history of logistic models).

Since then, one of the traditional domains of mathematical biology has focused on population dynamics. This field of research typically leverages methods from differential equations and dynamical systems to understand the size and features of biological populations.

The effectiveness of population dynamics has very often reached further out from its original targets and has provided extremely solid links with other branches of science and mathematics. For instance, the methods developed for problems in population dynamics frequently find applications in epidemiology (see, e.g., [85] for recent applications of logistic models to concrete descriptions of the COVID-19 pandemic). The nowadays very popular family of "SIR" models introduced by Kermack and McKendrick in epidemiology is a particular instance of population dynamics systems in which the population is divided into compartments: susceptibles, infected, and removed (see [58]).

Moreover, the questions posed in the setting of population dynamics often require the involvement of different mathematical specializations, such as game theory, control theory, optimization, etc. (see, e.g., [34, 52, 78, 79]).

The development of population dynamics has also deeply impacted social sciences and anthropology, providing quantitative settings to describe complicated social interactions, such as crime occurrences, migrations, foundations of political parties, voting processes, and circulation of ideas (see [12, 13, 81] and the references therein).

Finally, ideas from population dynamics have also been applied in the context of economics, predicting the phenomena of adoption of a new product or technology (the Bass model, see, e.g., [8, 54]), as well as knowledge diffusion in macroeconomics, and competition between old technologies and newer ones (Lotka–Volterra, see, e.g., [19, 98]).

1.1 Themes and Aims of This Work

Among the several models dealing with the dynamics of biological systems, the case of populations engaging into a mutual conflict seems to be unexplored. This work aims at laying the foundations of a new model describing two populations competing for the same resources with one aggressive population which may attack the other: Concretely, one may think of a situation in which two populations live together in the same territory and share the same environmental resources, till one population wants to prevail and try to overwhelm the other one.

We consider this situation as a "civil war," since the two populations share land and resources; the two populations may be equally fit to the environment (and, in this sense, they are "indistinguishable," up to the aggressive attitude of one of the populations), or they can have a different compatibility to the resources (in which case one may think that the conflict could be motivated by the different accessibility to environmental resources).

Given the lack of reliable data related to civil wars, a foundation of a solid mathematical theory for this type of conflicts may only leverage the deduction of the model from first principles: We follow this approach to obtain the description of the problem in terms of a system of two ordinary differential equations, each describing the evolution in time of the density of one of the two populations.

The method of analysis that we adopt is a combination of techniques from different fields, including ordinary differential equations, dynamical systems, and optimal control.

This viewpoint allows us to rigorously investigate the model, with a special focus on a number of mathematical features of concrete interest, such as the possible extinction of one of the two populations and the analysis of the strategies that lead to the victory of the aggressive population.

In particular, we analyze the *dynamics of the system*, characterizing the equilibria and their features (including possible basins of attraction) in terms of the different parameters of the model (such as relative fitness to the environment, aggressiveness, and effectiveness of strikes). Moreover, we study the initial configurations which may lead to the victory of the aggressive population, also taking into account different possible *strategies* to achieve the victory: Roughly speaking, we suppose that the aggressive population may adjust the parameter describing the aggressiveness in order to either dim or exacerbate the conflict with the aim of destroying the second population (of course, the war has a cost in terms of life for both the populations; hence the aggressive population must select the appropriate strategy in terms of the structural parameters of the system). We show that the initial data allowing the victory of the aggressive population does not exhaust the all space, namely *there exist initial configurations for which the aggressive population cannot make the other extinct*, regardless of the strategy adopted during the conflict.

Furthermore, *for identical populations with the same fit to the environment, the constant strategies suffice* for the aggressive population to possibly achieve the victory: Namely, if an initial configuration admits a piecewise continuous in time strategy that leads to the victory of the aggressive population, then it also admits a constant in time strategy that reaches the same objective (and of course, for the aggressive population, the possibility of focusing only on constant strategies would entail concrete practical advantages).

Conversely, *for populations with different fit to the environment the constant strategies do not exhaust all the winning strategies*: That is, in this case, there are initial conditions which allow the victory of the aggressive population only under the exploitation of a strategy that is not constant in time.

In any case, we also prove that *strategies with at most one jump discontinuity are sufficient* for the aggressive population: Namely, independently from the relative fit to the environment, if an initial condition allows the aggressive population to reach the victory through a piecewise continuous in time strategy, then the same goal can be reached using a "bang-bang" strategy with at most one jump.

We also discuss the *winning strategies that minimize the duration of the war*: In this case, we will show that jump discontinuous strategies may be not sufficient and interpolating arcs have to be taken into account.

1.2 Disclaimer

In no way do the authors of this book suggest that the model has implications of military or sociological type. This model is designed to expand the family of Lotka–Volterra systems by introducing a novel element, justifiable as an aggression term within the diverse interpretations of Lotka–Volterra systems—whether viewed through the lens of population dynamics or as models of business competition. Conversely, the incorporation of this new term significantly alters the system's behavior, thereby raising questions distinct from those prevalent in the existing literature, in particular from a control theory point of view. We suggest that our approach could be useful in other contexts. Through this book, our intention is to inspire and furnish an accessible resource for researchers possessing a robust mathematical background, particularly those intrigued by the development of competitive systems and models necessitating investigations into controllability.

1.3 Organization of This Monograph

In Chap. 2 we describe in detail the model starting from prime principles, and in Chap. 3 we present the main results of this monograph.

After having clarified the main notation used throughout this monograph in Chaps. 4 and 5, we will exploit methods from ordinary differential equations and dynamical systems to describe the equilibria of the system and their possible basins of attraction. The dependence of the dynamics on the structural parameters, such as fit to the environment, aggressiveness, and efficacy of attacks, is discussed in detail in Chap. 6.

Chapter 7 is devoted to the analysis of the strategies that allow the first population to eradicate the second one (this part needs an original combination of methods from dynamical systems and optimal control theory).

We conclude our work with a brief chapter offering guidance on classical bibliography. This section aims to assist readers seeking background on the theories we employ.

Description of the Model

2

We now describe in detail our model of conflict between the two populations and the attack strategies pursued by the aggressive population. We assume that two populations compete for the same resources; this leads to the standard competitive Lotka–Volterra system for their densities u and v, as introduced[1] in [65, 95], see also [66, 96].

We then incorporate the fact that one population—the one with density u—deliberately attacks the other. As a result, both populations suffer some losses.

The key point in our analysis is that the clashes do not depend on the chance of meeting between the two populations, given by the quantity uv, as it happens in many other works in the literature (starting from the publications of Lotka and Volterra, [66, 96]), but they are sought by the first population and only depend on the size u of the first population and on its level of aggressiveness a (or the portion of the population devoted to the attacks).

The resulting model is

$$\begin{cases} \dot{u} = u(1 - u - v) - acu, & \text{for } t > 0, \\ \dot{v} = \rho v(1 - u - v) - au, & \text{for } t > 0, \end{cases} \quad (2.1)$$

where a, c, and ρ are positive real numbers. Here, the coefficient ρ models the second population's fitness with respect to the first one when resources are abundant for both; it is linked with the exponential growth rate of the two species. The parameter c stands for the quotient of endured per inflicted damages for the first population. Deeper justifications to the model (2.1) will be given in Sect. 2.1. The complete description of the trajectories of the dynamical system (2.1) is presented in Sect. 3.1.

---

[1] This model was originally designed to describe a predator–prey system. Given its broad flexibility, it is often regarded as a paradigmatic model for competition.

© The Author(s), under exclusive license to Springer Nature Switzerland AG 2024
E. Affili et al., *A New Lotka-Volterra Model of Competition With Strategic Aggression*, Frontiers in Mathematics, https://doi.org/10.1007/978-3-031-67210-1_2

Notice that the size of the second population v may become negative in finite time, while the first population is still alive. The situation where $v = 0$ and $u > 0$ represents the extinction of the second population and the victory of the first one.

To describe our results, for communication convenience (and in spite of our personal fully pacifist beliefs), we take the perspective of the first population, that is, the aggressive one; the objective of this population is to overwhelm the other one, and, to achieve that, it can influence the system by tuning the parameter a.

From now on, we may refer to the parameter a as the *strategy*, which may also depend on time, and we will say that it is *winning* if it leads to the victory of the first population.

The main questions that we deal with in this monograph are:

1. The characterization of the *initial conditions for which there exists a winning strategy*
2. The *success of the constant strategies*, compared to all possible strategies
3. The *construction of a winning strategy* for a given initial datum
4. The *existence of a single winning strategy independently of the initial datum*

We discuss all these topics in Sect. 3.3, presenting concrete answers to each of these problems.

Also, since to our knowledge, this is the first time that system (2.1) is considered, in Sects. 3.1 and 3.2 we discuss the dynamics and some exciting results about the dependence of the basins of attraction on the other parameters.

It would also be extremely interesting to add the space component to our model, by considering a system of reaction–diffusion equations. This will be the subject of further work.

2.1 Motivations

The classic Lotka–Volterra equations for modeling predator–prey systems were first introduced independently in [65] and [95]. Later, the models were extended to other types of interaction between two populations, including competition (see [96]), and to model other phenomena involving competition, for example, in technology substitution [70]. The competitive Lotka–Volterra system concerns the sizes $u_1(t)$ and $u_2(t)$ of two species competing for the same resources. The system that the couple $(u_1(t), u_2(t))$ solves is

$$\begin{cases} \dot{u}_1 = r_1 u_1 \left(\sigma - \dfrac{u_1 + \alpha_{12} u_2}{k_1} \right), & t > 0, \\ \dot{u}_2 = r_2 u_2 \left(\sigma - \dfrac{u_2 + \alpha_{21} u_1}{k_2} \right), & t > 0, \end{cases} \qquad (2.2)$$

where $r_1, r_2, \sigma, \alpha_{12}, \alpha_{21}, k_1$, and k_2 are nonnegative real numbers.

2.1 Motivations

Here, the coefficients α_{12} and α_{21} represent the competition between individuals of different species, and indeed they appear multiplied by the term $u_1 u_2$, which represents a probability of meeting.

The coefficient r_i is the exponential growth rate of the i-th population, that is, the reproduction rate that is observed when the resources are abundant. The parameters k_i are called carrying capacity and represent the number of individuals of the i-th population that can be fed with the resources of the territory that are quantified by σ. It is however usual to renormalize the system in order to reduce the number of parameters. In general, u_1 and u_2 are normalized so that they vary in the interval $[0, 1]$, thus describing densities of populations.

The behavior of the system depends substantially on the values of $\alpha_{12}\frac{k_2}{k_1}$ and $\alpha_{21}\frac{k_1}{k_2}$ with respect to the threshold value 1, or simply on α_{12} and α_{21} if $k_1 = k_2$ (see, e.g., [9]). In this latter case, if $\alpha_{12} < 1 < \alpha_{21}$, then the first species u_1 has an advantage over the second one u_2 and will eventually prevail; if α_{12} and α_{21} are both strictly above 1, then the first population that penetrates the environment (that is, the one that has a greater size at the initial time) will persist, while the other will extinguish; if instead α_{12} and α_{21} are both equal or below 1, then an attractive coexistence equilibrium appears.

Some modifications of the Lotka–Volterra model were made in stochastic analysis by adding a noise term of the form $-f(t)u_i$ in the i-th equation, finding some emerging phenomena of phase transition, see, e.g., [48].

The ODE system (2.2) has been extended to study the case of two competitive populations that diffuse in space. Many different types of diffusion have been compared, and one can find a huge literature on the topic, see [25,68,74] for some examples and [72] for a more general overview. We point out that other dynamical systems exhibiting finite time extinction of one or more species living in some heterogeneous environments have been considered in the literature, see, for example, the model in [36] for the predator–prey behavior of cats and birds that has been thereafter widely studied.

In this monograph, we focus not only on *basic competition for resources*, but also on *situations of open hostility*. In social sciences, war models are in general little studied; indeed, the collection of data up to modern times is hard for the lack of reliable sources. Also, there is still much discussion about what factors are involved and how to quantify them: In general, the outcome of a war does not only depend on the availability of resources, but also on more subtle factors as the commitment of the population and the knowledge of the battlefield, see, e.g., [91]. Instead, the causes of war were investigated by the statistician L.F. Richardson, who proposed some models for predicting the beginning of a conflict, see [84].

In addition to the human populations, behavior of hostility between groups of the same species has been observed in chimpanzee. Other species with complex social behaviors are able to coordinate attacks against groups of different species: ants versus termites, agouti versus snakes, and small birds versus hawk and owls, see, e.g., [93].

The model that we present here is clearly a simplification of reality. Nevertheless, we try to capture some important features of conflicts between rational and strategic populations, introducing in the mathematical modeling the idea that a conflict may be sought and the parameters that influence its development may be conveniently adjusted.

Specifically, in our model, the interactions between populations are not merely driven by chance, but rather the strategic decisions of the population play a crucial role in the final outcome of the conflict, and we consider this perspective as an interesting novelty in the mathematical description of competitive environments.

At a technical level, our aim is to introduce a model for conflict between two populations u and v, starting from the model when the two populations compete for food and modifying it to add the information about the clashes. We imagine that each individual of the first population u decides to attack an individual of the second population with some probability a in a given period of time. As an outcome, the individual of the first population has a probability ζ_u of being killed and a probability ζ_v of killing one opponent. One may think that hostilities take the form of "duels," that is, one-to-one fights, whose single outcome does not depend on the total number of individuals of the populations. Notice that in some duel the fighters might both be killed. Thus, after one time period, the casualties for the first and second populations are $a\zeta_u u$ and $a\zeta_v u$, respectively. The same conclusions are found if we imagine that the first population forms an army to attack the second, which tries to resist by recruiting an army of proportional size. At the end of each battle, a ratio of the total soldiers is dead, and this is again of the form $a\zeta_u u$ for the first population and $a\zeta_v u$ for the second one.

Another effect that we want to take into account is the drop in the fertility of the population during wars. This seems due to the fact that families suffer some income loss during war time, because of a lowering of the average productivity and lacking salaries only partially compensated by the state; another reason possibly discouraging couples to have children is the increased chance of death of the parents during war. As pointed out in [94], in some cases the number of lost births during wars is comparable to the number of casualties. However, it is not reasonable to think that this information should be included in the exponential growth rates r_u and r_v, because the fertility drop really depends on the intensity of the war. For this reason, we introduce a population loss rate for u and v given by $c_u au$ and $c_v au$, respectively, where $c_u \geqslant 0$ and $c_v \geqslant 0$ are given parameters.

Finally, for simplicity, we also suppose that the clashes take place apart from inhabited zone, without having influence on the harvesting of resources.

2.2 The Notion of Aggressiveness

Concerning the notion of "aggressiveness," we remark that obviously in our simplified model this term has merely a mathematical meaning, and not a social, psychological, or legal connotation.

From the historical point of view, the notion of "aggression" in relation to military actions was probably formalized for the first time in 1919, on the occasion of the Treaty of Versailles (Article 231, often referred to as the "War Guilt Clause," stated that "The Allied

2.2 The Notion of Aggressiveness

and Associated Governments affirm and Germany accepts the responsibility of Germany and her allies for causing all the loss and damage to which the Allied and Associated Governments and their nationals have been subjected as a consequence of the war imposed upon them by the *aggression* of Germany and her allies").

Similar clauses were also used in the Treaty of Saint-Germain-en-Laye (1919), in the Treaty of Neuilly (1919), in the Treaty of Trianon (1920), and in the Treaty of Sévres (1920). Articles of this type have been used as legal bases to extract money reparations for the war's devastations and costs. Notwithstanding the existence of an International Criminal Court, the definition for a war of aggression is not univocal and it is often controversial.

The notion of "aggressiveness" has also been commonly employed in connection with wars in several historical contexts. For instance, in relation to ancient empires, the word "aggressive" has also been very often adopted by scholars (e.g., "Roman aggression" [82, page 229], as well as "...the most aggressive ancient and modern civilized states.", see [56, page 33]; "the populous and aggressive Parthian (Persian) Empire," see [56, page 76]; also about ancient Egypt "In the Old Kingdom, warfare was supposed to be aggressive" [82, page 77]; in relation to the Hellenistic World "aggressive kings such as Philip and Alexander" [82, page 171]; and, with regard to Eastern imperial dynasties, "the Japanese court faced the aggressive Tang Empire" [82, page 52]; as regards the prehistoric and pre-Columbian Mesoamerican societies, "the major aggressor was the Culhua-Mexica (or Aztec) empire, which rather easily overran the area in 1486 and in 1506," see [2, page 340]; "Aztec aggression," see [2, page 357]; "It is difficult to say exactly What the statuittle doubt that it was a politically weak, militarily aggressive, and probably tributary group," see [2, page 398]; "The Warrau appear to have been pushed into very marginal swamp areas by their notoriously expansive and aggressive neighbors, the Caribs and Arawaks" [63, page 203]).

In this context, there are also classical examples of "peaceful" populations (such as "the certain peaceful Inuit groups, the Semai, and the La Paz Zapotec, as well as on several other peaceful cultures," see [61, page 721]).

However, aggressiveness and peacefulness may vary within different communities of the same population (e.g., "it would be incorrect to generalize that Zapotec culture overall has a low level of aggression based solely on data from the peaceful La Paz community, or conversely, to generalize on the basis of fieldwork in a different Zapotec community than La Paz that all Zapotec communities are violent. Jean Briggs voices a similar caution that not all Inuit bands are as peaceful as the groups she describes," see [61, page 727]).

Also in cases of Indian massacres in North America, the role of the "aggressive" or "colonialistic" population has also often emerged quite clearly (in fact, scholars speak about "colonial aggression," see, e.g., [3, pages 133 and 247]), though we have also occurrences of aggressive behaviors of indigenous populations (see [56, pages 128–129] in the context of "Apache-Navajo aggressiveness" and "the aggressive Mohave," see also [63, page 58] for "The Comanche of the Southern Plains and the Yanomama have been described as particularly aggressive").

The distinction between aggressive and peaceful populations is also a notion adopted by scholars (e.g., "the aggressive groups acquired territory at the expense of more passive ones," see [56, page 129]; "regions and periods of frequent bitter warfare are often centered on especially aggressive societies that spoil their neighborhood," see [56, page 177]; "the Semai [...] tradition of flight from violence is a consequence of countless defeats and slave raiding at the hands of the more numerous and aggressive Malays. In other words, the Sexnai can be characterized as defeated refugees," see [56, page 206]).

Obviously, aggressive treats may also change in time and according to circumstances (e.g., "The hyperaggressive Norsemen have become the pacific Scandinavians," see [56, page 130]).

The strict link between the notions of "war" and "aggression" is at the basis of [75].

Of course, the "aggressiveness" of a population is sometimes highly influenced by its political leader (see, e.g., [56, page 175] for "Napoleon's aggressive use of" the French Revolution). Moreover, in cases of national, ethnical, racial, or religious genocides, the role of an "aggressive" population is usually very apparent.

Aggressiveness is not only found in humans but is a well-studied behavior in the animal realm. Konrad Lorenz, one of the founders of ethology, also wrote an influential book fully dedicated on the topic of "aggression," see [64]. In his opinion, "the aggression of so many animals towards members of their own species is in no way detrimental to the species but, on the contrary, is essential for its preservation" [64]. The same concepts were also confirmed and presented by Dawkins in his influential book "The selfish gene" [27]. Aggression is a complex behavior that can manifest in a variety of ways, including physical attacks, threats, and dominance displays.

Aggression is in fact observed in a wide range of animal species, from insects to fish, birds, and mammals [26]. Interspecific aggressive behavior seems to be crucial in order to defend territory, protect the offspring, and establish dominance hierarchies that ensure breeding rights to the triumphing male. In ecosystems around the world, top predators kill, harass, and steal food from smaller predators. These direct, aggressive interactions, generally referred to as interference competition, are widespread and substantial and can have profound consequences for the distributions and population dynamics of smaller predators. These patterns of suppression and coexistence vary across systems and species, see [87].

Aggressive business policies, including predatory pricing, excessive discounting, and exclusive dealing, are tactics employed by companies to gain or maintain a competitive advantage in the market. These practices can harm consumers and rival businesses, and they are often subject to legal scrutiny [6]. Many retailers use excessive discounting as a promotional tactic to attract customers during periods of low demand. However, this practice can harm smaller competitors and lead to higher prices in the long run. Some authors refer to this strategy as "price wars," see [40].

Moreover, with a slight abuse of notation, it is fascinating to include in the study of "civil wars" possibly the most ancient human conflict on a large scale, namely the long war of attrition between the individuals of *Homo sapiens* and those of *Homo neanderthalensis*, caused by the expansion of the sapiens out of Africa about 60 or 70 thousand years ago, which led to the extinction of the Neanderthals around 40 thousand years ago, with a concrete overlap of the two species for between about 2 and 5 thousand years, see [41].

On the one hand, this conflict cannot be classified as a civil war in the modern sense of the term, also because neanderthalensis and sapiens are recognized as two separate species. On the other hand, the two species present a strikingly similar anatomy and share 99.7% of DNA, and there is even strong indication of interbreeding, see [38].

Both species were certainly acquainted with war actions: For instance, signs of warfare are typically considered skull traumas and parry fractures, which seem to be especially common in young males, see [73].

The reasons for the supremacy of the sapiens species are still under intense debate, and they may include a refined symbolic intelligence, a more articulated language, the adoption of superior ranged weapons, more advanced social systems, a more specialized division of labor, as well as possibly a more aggressive and better organized expansion of the sapiens which broke the preexisting demographic balance, see, e.g., [51, 60]. Let us also mention the mathematical modeling of Sapiens–Neanderthal interaction proposed in [33] where the author adopts the point of view of considering pure quadratic competition between the two species and explicitly neglects the war perspective.

2.3 Derivation of the Model

Now we derive the system of equations from an heuristic analysis. As in the Lotka–Volterra model, it is assumed that the change of the size of the population in an interval of time Δt is proportional to the size of the population $u(t)$, that is,

$$u(t + \Delta t) - u(t) \approx u(t) f(u, v)$$

for some appropriate function $f(u, v)$. In particular, $f(u, v)$ should depend on resources that are available and reachable for the population. The maximum number of individuals that can be fed with all the resources of the environment is k; taking into account all the individuals of the two populations, the available resources are

$$k - u - v.$$

Notice that we suppose here that each individual consumes the same amount of resources, independently of its belonging. In our model, this assumption is reasonable since all the individuals belong to the same species. Also, the competition for the resources only depends on the number of individuals, independently on their identity.

Furthermore, our model is sufficiently general to take into account the fact that the growth rate of the populations can be possibly different. In practice, this possible difference could be the outcome of a cultural distinction, or it may be also due to some slight genetic differentiation, as it happened in the case of Homo Sapiens and Neanderthal mentioned in the previous section.

Let us call r_u and r_v the fertility of the first and second populations, respectively. The contribution to the population growth rate is given by

$$f(u, v) := r_u \left(1 - \frac{u+v}{k}\right),$$

and these effects can be comprised in a typical Lotka–Volterra system.

Instead, in our model, we also take into account the possible death rate due to casualties. In this way, we obtain a term such as $-a\zeta_u$ to be added to $f(u, v)$. The fertility losses give another term $-ac_u$ for the first population. We also perform the same analysis for the second population, with the appropriate coefficients.

With these considerations, the system of the equations that we obtain is

$$\begin{cases} \dot{u} = r_u u \left(1 - \dfrac{u+v}{k}\right) - a(c_u + \zeta_u)u, & t > 0, \\ \dot{v} = r_v v \left(1 - \dfrac{v+u}{k}\right) - a(c_v + \zeta_v)u, & t > 0. \end{cases} \quad (2.3)$$

As usual in these kinds of models, we can rescale the variables and the coefficients in order to find an equivalent model with fewer parameters.

Hence, we perform the changes of variables

$$\tilde{u}(\tilde{t}) = \frac{u(t)}{k}, \quad \tilde{v}(\tilde{t}) = \frac{v(t)}{k}, \quad \text{where} \quad \tilde{t} = r_u t,$$

$$\tilde{a} = \frac{a(c_v + \zeta_v)}{r_u}, \quad \tilde{c} = \frac{c_u + \zeta_u}{c_v + \zeta_v} \quad \text{and} \quad \rho = \frac{r_v}{r_u}, \quad (2.4)$$

and, dropping the tildas for the sake of readability, we finally get the system in (2.1). We will also refer to it as the civil war model.

From the change of variables in (2.4), we notice in particular that a may now take values in $[0, +\infty)$.

2.4 Interpretation of the Model in an Economic Key

The competitive Lotka–Volterra system is already used to study some market phenomena as technology substitution, see, e.g., [14,70,99], and our model aims at adding new features to such models.

Concretely, in the technological competition model, one can think that u and v represent the capitals of two companies, producing for instance computers, or cell phones, etc. In this setting, to start with, one can suppose that the first company produces a very successful

2.4 Interpretation of the Model in an Economic Key

product, say computers with a certain operating system, in an infinite market, reinvesting a proportion r_u of the profits into the production of additional items, which are purchased by the market, and so on: In this way, one obtains a linear equation of the type $\dot{u} = r_u u$, with exponentially growing solutions. The case in which the market is not infinite, but admits a saturation income threshold k, would correspond to the equation

$$\dot{u} = r_u u \left(1 - \frac{u}{k}\right).$$

Then, when a second computer company comes into the business, selling computers with a different operating system to the same market, one obtains the competitive system of equations

$$\begin{cases} \dot{u} = r_u u \left(1 - \frac{u+v}{k}\right), \\ \dot{v} = r_v v \left(1 - \frac{v+u}{k}\right). \end{cases} \tag{2.5}$$

At this stage, the first company may decide to use an "aggressive" strategy in order to harm the rival company and set it out of the market (for instance through the spreading of a virus attacking the other company's operating system, or by some marketing campaigns). Once the competition of the second company is removed, the first company can then exploit the market in a monopolistic regime. To model this strategy, one can suppose that the first company invests a proportion of its capital in the project and diffusion of the virus, according to a quantifying parameter $a_u \geqslant 0$, thus producing the equation

$$\dot{u} = r_u u \left(1 - \frac{u+v}{k}\right) - a_u u. \tag{2.6}$$

This directly impacts the capital of the second company proportionally to the virus spread, since the second company has to spend money to project and release antiviruses, as well as to repay unsatisfied customers, hence resulting in a second equation of the form

$$\dot{v} = r_v v \left(1 - \frac{v+u}{k}\right) - a_v u. \tag{2.7}$$

The case $a_u = a_v$ would correspond to an "even" effect in which the costs of producing the virus are in balance with the damages that it causes. It is also realistic to take into account the case $a_u < a_v$ (e.g., the first company manages to produce and diffuse the virus at low cost, with high impact on the functionality of the operating system of the second company) as well as the case $a_u > a_v$ (e.g., the cost of producing and diffusing the virus is high with respect to the damages caused).

We remark that Eqs. (2.6) and (2.7) can be set into the form (2.3), thus showing the interesting versatility of our model also in financial mathematics.

Even the original Bass model [8], introduced to describe the number of adopters of a durable good, can be extended to the case of two products competing for the same market, leading to a competitive system of the Lotka–Volterra type, see, e.g., [15, 59]. In such a framework, if we neglect "innovators," new people adopt a product by imitation, namely, calling u and v the portions of adopters of the two goods, their rate of change is proportional to the portion of people that do not adopt neither product, i.e., $(1 - u - v)$, times the portion that already adopted the product, i.e., u or v. The system then takes the form (2.5) with $k = 1$. One can then envision the fact that one of the two companies producing the goods resorts to some type of aggressive marketing policy in order to harm the rival. Without describing the specific mechanisms of such a policy, we just make the general assumption that, on one hand, it requires some consumption of resources by the company, and, on the other hand, it produces some damages to the rival one; the outcome is the reduction of the rates of change of adopters of the two products, and we assume that these are proportional. Under these assumptions, we end up with the system (2.6)–(2.7).

Finally, it is natural to envision other contexts where our model could be pertinent, such as dynamic games. Let us mention for instance real-time strategy computer games that combine resource management and war confrontation, see [76].

As a final comment, let us stress that certainly our model does not aim to capture all the complexity of the phenomena intertwined with civil wars, and other mathematical approaches to the problem can certainly be very beneficial for a deeper understanding of the problem. Other possible tools of investigations naturally include (but are not limited to) kinetic models and Boltzmann-type equations, mean-field games, agent-based models, and active particles methods, see, e.g., [1, 4, 10, 11, 32, 50, 62] and the references therein.

Thus, our objective here is just to propose a simple, stylized model to describe such a complex scenario as the interaction between rival species and communities. The outcomes of our approach are that, on one hand, it allows us for a rigorous analytic investigation of the mathematical model proposed. On the other hand, we believe that our results may capture some qualitative features of the real phenomenon under study, as we are now going to showcase in some detail.

Statement of the Main Results

We now describe in some technical detail the results that we obtain on the civil war model.

3.1 Basic Results on the Dynamics

We denote by $(u(t), v(t))$ a solution of (2.1) starting from a point

$$(u(0), v(0)) \in [0, 1] \times [0, 1].$$

We will also refer to the *orbit* of $(u(0), v(0))$ as the following subset of $\mathbb{R}^2$:

$$\{(u(t), v(t)) : t \in \mathbb{R}\},$$

thus both positive and negative times, while the *trajectory* (referred to in some texts as simply *solution* or *phase curve through the point* $(u(0), v(0))$ at time $t = 0$, see [46, 100]) is the set

$$\{(u(t), v(t)) : t \geqslant 0\}.$$

As already mentioned in the discussion below formula (2.1), v can reach the value 0 and even negative values in finite time. However, we suppose that the dynamics stops when the value $v = 0$ is reached for the first time. At this point, the conflict ends with the victory of the first population u, which can continue its evolution with a classical Lotka–Volterra equation of the form

$$\dot{u} = u(1 - u)$$

and which would certainly fall into the attractive equilibrium $u = 1$. The only other possibility is that the solution remains in the set $[0, 1] \times (0, 1]$ for all times. Indeed, on the rest of the boundary of this square, there hold

$$u(t) = 0 \implies \dot{u}(t) = 0,$$
$$u(t) = 1 \implies \dot{u}(t) \leqslant 0,$$
$$\text{and} \quad v(t) = 1 \implies \dot{v}(t) \leqslant 0.$$

Remark 3.1 In a nutshell, for any solution with initial datum $(u(0), v(0)) \in [0, 1] \times [0, 1]$, one of the following situations occurs:

(1) $(u(t), v(t)) \in [0, 1] \times [0, 1]$ for all $t \geqslant 0$.
(2) There exists a unique $T \geqslant 0$ such that $v(T) = 0$, $u(T) > 0$ and $(u(t), v(t)) \in [0, 1] \times (0, 1]$ for all $t \in [0, T)$.

Owing to this dichotomy, we define the *stopping time* of the solution $(u(t), v(t))$ as

$$T_s(u(0), v(0)) := \begin{cases} +\infty & \text{if situation (1) occurs,} \\ T & \text{if situation (2) occurs.} \end{cases} \tag{3.1}$$

From now on, we will implicitly consider solutions $(u(t), v(t))$ only for $t \in [0, T_s(u(0), v(0)))$.

We now specify the class of admissible strategies $a(t)$. In view of the applications, one is led to allow $a(t)$ to be nonconstant and discontinuous, so we consider the following class of admissible strategies:

$$\mathcal{A} := \big\{ a : [0, +\infty) \to [0, +\infty) \text{ s.t. } a \text{ is continuous} \tag{3.2}$$
$$\text{except at most at a finite number of points} \big\}.$$

A *solution related to a strategy* $a(t) \in \mathcal{A}$ is a pair

$$(u(t), v(t)) \in C^0((0, +\infty)) \times C^0((0, +\infty)),$$

which is C^1 outside the points of discontinuity of $a(t)$ and solves the system (2.1) outside these points. Moreover, once the initial datum is imposed, the solution is assumed to be continuous up to $t = 0$. The existence and uniqueness of solutions in this setting is standard: One just considers a juxtaposition of classical Cauchy problems starting at the times coinciding with the discontinuity points of a.

We then analyze the dynamics of (2.1) with a particular emphasis on possible strategies. To do this, we consider the *basin of attraction* of the equilibrium $(0, 1)$, i.e.,

3.2 Constant Strategies

$$\mathcal{B} := \Big\{ (u(0), v(0)) \in [0, 1] \times [0, 1] \text{ s.t.} \qquad (3.3)$$
$$T_s(u(0), v(0)) = +\infty, \ (u(t), v(t)) \stackrel{t \to \infty}{\longrightarrow} (0, 1) \Big\},$$

which corresponds to the set of the initial points for which the first population gets extinct (in infinite time) and the second one survives.

Furthermore, we set

$$\mathcal{E} := \Big\{ (u(0), v(0)) \in [0, 1] \times [0, 1] \qquad (3.4)$$
$$\text{s.t. } T_s(u(0), v(0)) < +\infty \Big\},$$

namely the set of initial points for which we eventually have the victory of the first population and the extinction of the second one.

Of course, the sets $\mathcal{B}$ and $\mathcal{E}$ depend on the parameters a, c, and ρ; we will express this dependence by writing $\mathcal{B}(a, c, \rho)$ and $\mathcal{E}(a, c, \rho)$ when it is needed and omit it otherwise for the sake of readability. The dependence on parameters will be carefully studied in Chap. 6.

3.2 Constant Strategies

The first step toward the understanding of the dynamics of (2.1) consists in the analysis of the behavior of the system for constant coefficients.

To this end, we introduce some notation. Following the terminology of [100, Section 1.1.], we say that an equilibrium point (or fixed point) of the dynamics is a (hyperbolic) *sink* if all the eigenvalues of the linearized map have strictly negative real parts, a (hyperbolic) *source* if all the eigenvalues of the linearized map have strictly positive real parts, and a (hyperbolic) *saddle* if some of the eigenvalues of the linearized map have strictly positive real parts and some have strictly negative real parts (since in this monograph we work in dimension 2, this means that one eigenvalue has positive real part and the other has negative real part).

We also recall that sinks are asymptotically stable (and sources are asymptotically stable for the reversed-time dynamics), see, e.g., [100, Theorem 1.1.1].

With this terminology, we state the following theorem:

Theorem 3.2 (Dynamics of System (2.1)**)** *For given positive constants a, ρ, and c, the system* (2.1) *has the following features:*

(i) When $0 < ac < 1$, there are three equilibria: $(0, 0)$ is a source, $(0, 1)$ is a sink, and

$$(u_s, v_s) := \left(\frac{1-ac}{1+\rho c} \rho c, \frac{1-ac}{1+\rho c} \right) \in (0, 1) \times (0, 1) \tag{3.5}$$

is a saddle.
(ii) When $ac > 1$, there are two equilibria: $(0, 1)$ is a sink and $(0, 0)$ is a saddle.
(iii) When $ac = 1$, there are two equilibria: $(0, 1)$ is a sink and $(0, 0)$ corresponds to a positive eigenvalue and a null one.
(iv) We have

$$[0, 1] \times [0, 1] = \mathcal{B} \cup \mathcal{E} \cup \mathcal{M}, \tag{3.6}$$

where $\mathcal{B}$ and $\mathcal{E}$ are defined in (3.3) and (3.4), respectively, and $\mathcal{M}$ is a smooth curve.
(v) There holds that:

- *$\mathcal{M}$ is the stable manifold of (u_s, v_s) if $0 < ac < 1$.*
- *$\mathcal{M}$ is the center manifold of $(0, 0)$ if $ac = 1$.*
- *$\mathcal{M}$ is the stable manifold of $(0, 0)$ if $ac > 1$.*

Moreover, trajectories starting in $\mathcal{M}$ remain in $\mathcal{M}$ and converge to (u_s, v_s) if $0 < ac < 1$ and to $(0, 0)$ if $ac \geqslant 1$ as t goes to $+\infty$.

Figure 3.1 depicts the different dynamics in the cases $ac < 1$ and $ac > 1$.

In the case $ac \neq 1$, i.e., when $\mathcal{M}$ is the stable manifold of a saddle point, the properties of $\mathcal{M}$ stated in Theorem 3.2 follow from the general theory of dynamical systems (see, e.g., [77]).

Instead, the case $ac = 1$ needs a special treatment, due to the degeneracy of one eigenvalue, and an ad hoc argument will be exploited to show that also in this degenerate case orbits starting in $\mathcal{M}$ are asymptotic to $(0, 0)$ in the future.

As a matter of fact, $\mathcal{M}$ acts as a dividing wall between the two basins of attraction, as described in (iv) of Theorem 3.2 and in the forthcoming Proposition 5.9.

Moreover, in the forthcoming Propositions 5.1 and 5.7, we will show that $\mathcal{M}$ can be written as the graph of a function. This is particularly useful because, by studying the properties of this function, we gain relevant pieces of information on the sets $\mathcal{B}$ and $\mathcal{E}$ in (3.3) and (3.4).

We point out that in Theorem 3.2 we find that the set of initial data $[0, 1] \times [0, 1]$ splits into three parts: the set $\mathcal{E}$, given in (3.4), made of points going to the extinction of the second population in finite time, the set $\mathcal{B}$, given in (3.3), which is the basin of attraction of the equilibrium $(0, 1)$, and the set $\mathcal{M}$, which is a manifold of dimension 1 that separates $\mathcal{B}$ from $\mathcal{E}$.

In particular, Theorem 3.2 shows that, also for our model, the Gause principle of exclusion is respected; that is, in general, two competing populations cannot (stably) coexist in the same territory, see, e.g., [30].

3.2 Constant Strategies

Fig. 3.1 The figures show a phase portrait for the indicated values of the coefficients. Plotted in black are the orbits of the points. The red dots represent the equilibria. The light blue region corresponds to $\mathcal{E}$, while the white region corresponds to $\mathcal{B}$

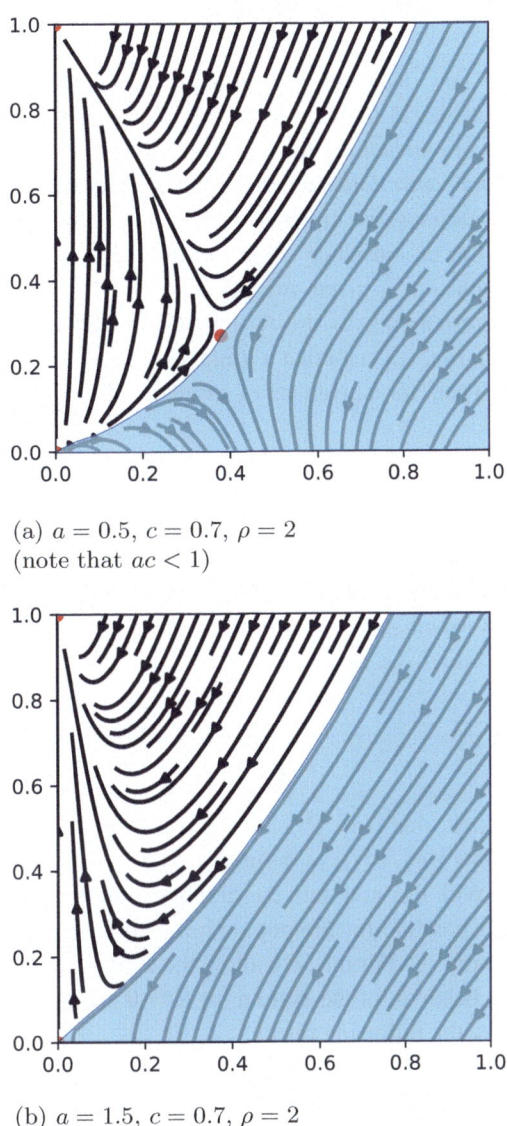

(a) $a = 0.5$, $c = 0.7$, $\rho = 2$ (note that $ac < 1$)

(b) $a = 1.5$, $c = 0.7$, $\rho = 2$ (note that $ac > 1$)

One peculiar feature of our system is that if the aggressiveness is too strong, the equilibrium $(0, 0)$ changes its "stability" properties, passing from a source (as in (i) of Theorem 3.2) to a saddle point (as in (ii) of Theorem 3.2). This shows that the war may have self-destructive outcomes; therefore, it is important for the first population to analyze the situation in order to choose a proper level of aggressiveness.

3.3 Winning Strategies

We now deal with the problem of choosing the strategy a, in the set of all admissible strategies $\mathcal{A}$ (defined by (3.2)), such that the first population wins, which is a problem of *target reachability* for a control-affine system (see also [16]). As we will see, the problem is not *controllable*, meaning that, starting from a given initial point, it is not always possible to reach a given target.

Let us introduce some terminology, which will be employed throughout this monograph. For a given strategy $a(\cdot) \in \mathcal{A}$, we let $\mathcal{E}(a(\cdot))$ denote the set defined by (3.4) corresponding to the system (2.1) with $a \equiv a(t)$; namely, this is the set of initial data (u_0, v_0) such that $T_s(u_0, v_0) < +\infty$ for the strategy $a(\cdot)$.

Then, for a given set of strategies $\mathcal{T} \subset \mathcal{A}$, we set

$$\mathcal{V}_\mathcal{T} := \bigcup_{a(\cdot) \in \mathcal{T}} \mathcal{E}(a(\cdot)), \qquad (3.7)$$

which represents the set of initial conditions for which u is able to win by choosing a suitable strategy in $\mathcal{T}$; we call $\mathcal{V}_\mathcal{T}$ the *victory set* with strategies in $\mathcal{T}$. We also say that $a(\cdot)$ is a *winning strategy* for the point (u_0, v_0) if $(u_0, v_0) \in \mathcal{E}(a(\cdot))$.

Moreover, we set

$$(u_s^0, v_s^0) := \left(\frac{\rho c}{1 + \rho c}, \frac{1}{1 + \rho c} \right). \qquad (3.8)$$

Notice that (u_s^0, v_s^0) is the limit point as $a \to 0$ of the sequence of saddle points $\{(u_s^a, v_s^a)\}_{a>0}$ defined in (3.5).

With this notation, the first question that we address is for which initial configurations it is possible for the population u to have a winning strategy, that is, to characterize the victory set. For this, we allow the strategy to take all the values in $[0, +\infty)$ (the details on the behavior of the system for $a = 0$ are contained in Proposition 6.3). In this setting, we have the following result:

Theorem 3.3 *We have:*

(i) For $\rho = 1$, we have that

$$\mathcal{V}_\mathcal{A} = \left\{ (u, v) \in [0, 1] \times [0, 1] \text{ s.t. } v - \frac{u}{c} < 0 \right\}. \qquad (3.9)$$

(ii) For $\rho < 1$, we have that

3.3 Winning Strategies

$$\mathcal{V}_A = \Big\{(u,v) \in [0,1]\times[0,1] \text{ s.t.}$$

$$v < \gamma_0(u) \text{ if } u \in [0, u_s^0], \tag{3.10}$$

$$v < \frac{u}{c} + \frac{1-\rho}{1+\rho c} \text{ if } u \in \left(u_s^0, \frac{\rho c(c+1)}{1+\rho c}\right]\Big\},$$

where

$$\gamma_0(u) := \frac{v_s^0}{(u_s^0)^\rho} u^\rho. \tag{3.11}$$

(iii) For $\rho > 1$, we have that

$$\mathcal{V}_A = \Big\{(u,v) \in [0,1]\times[0,1] \text{ s.t.}$$

$$v < \frac{u}{c} \text{ if } u \in [0, u_\infty], \tag{3.12}$$

$$v < \zeta(u) \text{ if } u \in \left(u_\infty, \frac{c}{(c+1)^{\frac{\rho-1}{\rho}}}\right]\Big\},$$

where

$$u_\infty := \frac{c}{c+1} \quad \text{and} \quad \zeta(u) := \frac{u^\rho}{c u_\infty^{\rho-1}}. \tag{3.13}$$

With this result in hand, it is natural to wonder whether the scenario changes if one restricts to constant strategies. Indeed, in practice, these are certainly easier to implement. The next result addresses this problem by showing that when $\rho = 1$ constant strategies are as good as all strategies, but instead when $\rho \neq 1$ victory cannot be achieved by only exploiting constant strategies:

Theorem 3.4 *Let $\mathcal{K} \subset \mathcal{A}$ be the set of constant functions. Then the following holds:*

(i) For $\rho = 1$, we have that $\mathcal{V}_A = \mathcal{V}_\mathcal{K} = \mathcal{E}(a)$ for any $a > 0$.
(ii) For $\rho \neq 1$, we have that $\mathcal{V}_\mathcal{K} \subsetneq \mathcal{V}_A$.

The result of Theorem 3.4, part (i), reveals a special rigidity of the case $\rho = 1$ in which, no matter which strategy u chooses, the victory depends only on the initial conditions, but it is independent of the strategy $a(t)$.

Instead, as stated in Theorem 3.4, part (ii), for $\rho \neq 1$ the choice of $a(t)$ plays a crucial role in determining which population is going to win, and constant strategies do not exhaust all the possible winning strategies.

Roughly speaking, when $\rho = 1$ constant strategies suffice to detect all possible winning configurations, while when $\rho \neq 1$ nonconstant strategies are necessary to detect all winning configurations.

We stress that $\rho = 1$ plays also a special role in the biological interpretation of the model, since in this case the two populations have the same fit to the environmental resource, and hence, in a sense, they are indistinguishable, up to the possible aggressive behavior of the first population.

Next, we show that the set $\mathcal{V}_\mathcal{A}$ can be recovered if we use piecewise constant functions with at most one discontinuity, which we call Heaviside functions.

Theorem 3.5 *There holds that $\mathcal{V}_\mathcal{A} = \mathcal{V}_\mathcal{H}$, where $\mathcal{H}$ is the set of Heaviside functions.*

In proving Theorem 3.5 we will actually answer to the third question mentioned in the Introduction section: For each point in $\mathcal{V}_\mathcal{A}$, we either have a constant winning strategy or a winning strategy of the type

$$a(t) = \begin{cases} a_1 & \text{if } t < T, \\ a_2 & \text{if } t \geq T, \end{cases}$$

for a suitable $T \in (0, T_s)$ and for a_i very small and a_j very large, the values of i and j depending on ρ. Our construction also enlightens the fact that the choice of the strategy depends on the initial datum, answering to the fourth question as well.

It is interesting to observe that the winning strategy that switches abruptly from a small to a large value could be considered, in the optimal control terminology, as a "bang–bang" strategy. Even in a target reachability problem, the structure predicted by Pontryagin's Maximum Principle is brought in light: The bounds of the set $\mathcal{V}_\mathcal{A}$, as given in Theorem 3.3, depend on the bounds that we impose on the strategy, that are, $a \in [0, +\infty)$.

It is natural to consider also the case in which the level of aggressiveness is constrained between a minimal and a maximal threshold, which corresponds to imposing $a \in [m, M]$ for given $0 \leq m \leq M \leq +\infty$, with $M > 0$. In this setting, we denote by $\mathcal{A}_{m,M}$ the class of piecewise continuous strategies $a(\cdot)$ in $\mathcal{A}$ such that $m \leq a(t) \leq M$ for all $t > 0$, and we call

$$\mathcal{V}_{m,M} := \mathcal{V}_{\mathcal{A}_{m,M}} = \bigcup_{\substack{a(\cdot) \in \mathcal{A} \\ m \leq a(t) \leq M}} \mathcal{E}(a(\cdot)). \tag{3.14}$$

Observe that in the case $M = +\infty$, the strategy actually satisfies $m \leq a(t) < +\infty$ since $a(\cdot) \in \mathcal{A}$. Then we have the following:

Theorem 3.6 *Let M and m be two real numbers such that $0 \leqslant m \leqslant M \leqslant +\infty$ with $M > 0$ and either $m \neq 0$ or $M \neq +\infty$. Then, for $\rho \neq 1$, we have the strict inclusion*

$$\mathcal{V}_{m,M} \subsetneq \mathcal{V}_A.$$

Notice that for $\rho = 1$, Theorem 3.4 gives instead that $\mathcal{V}_{m,M} = \mathcal{V}_A$.

3.4 Time Minimizing Strategy

Once established that it is possible to win starting at a certain initial condition, we are interested in knowing which of the possible strategies is best to choose. One condition that may be taken into account is the duration of the war. Now, this question can be written as a minimization problem with a proper functional to minimize, and therefore the classical Pontryagin theory applies.

To state our next result, we consider a given $(u_0, v_0) \in \mathcal{V}_{m,M}$ and, recalling the setting in (3.14), we define

$$\mathcal{S}(u_0, v_0) := \Big\{ a(\cdot) \in \mathcal{A}_{m,M} \text{ s.t. } (u_0, v_0) \in \mathcal{E}(a(\cdot)) \Big\}.$$

This is the set of all bounded strategies for which the trajectory starting at (u_0, v_0) leads to the victory of the first population.

To each $a(\cdot) \in \mathcal{S}(u_0, v_0)$, we associate the stopping time defined in (3.1), and we express its dependence on $a(\cdot)$ by writing $T_s(a(\cdot))$.

In this setting, we provide the following statement concerning the strategy leading to the quickest possible victory for the first population:

Theorem 3.7 *Given a point $(u_0, v_0) \in \mathcal{V}_{m,M}$, there exists a winning strategy $\tilde{a}(t) \in \mathcal{S}(u_0, v_0)$ for which*

$$T_s(\tilde{a}(\cdot)) = \min_{a(\cdot) \in \mathcal{S}} T_s(a(\cdot)).$$

Moreover, the optimal strategy satisfies

$$\tilde{a}(t) \in \{m, \, M, \, a_s(t)\},$$

where

$$a_s(t) := \frac{(1 - \tilde{u}(t) - \tilde{v}(t))[\tilde{u}(t)\,(2c + 1 - \rho c) + \rho c]}{\tilde{u}(t)\,2c(c+1)}, \qquad (3.15)$$

and $(\tilde{u}(t), \tilde{v}(t))$ is the trajectory emerging from (u_0, v_0) associated with $\tilde{a}(t)$.

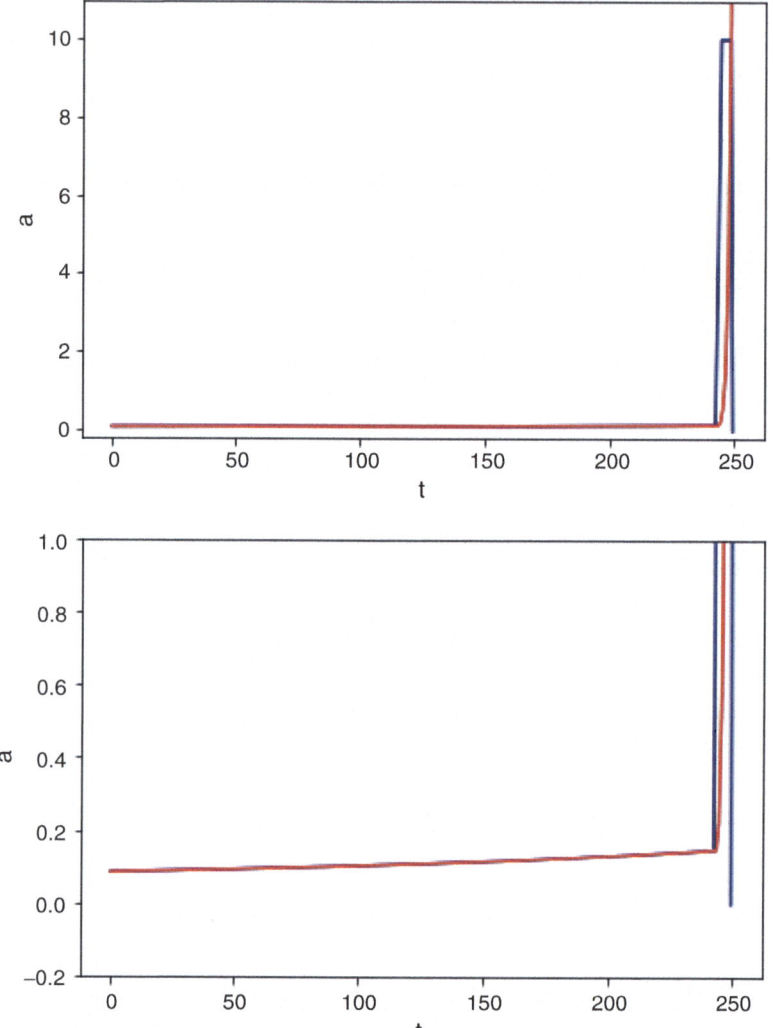

Fig. 3.2 The figure shows the result of a numerical simulation searching a minimizing time strategy $\tilde{a}(t)$ for the problem starting in $(0.5, 0.1875)$ for the parameters $\rho = 0.5$, $c = 4.0$, $m = 0$, and $M = 10$. Plotted in blue is the value found for $\tilde{a}(t)$ and in red the value of $a_S(t)$ for the corresponding trajectory $(u(t), v(t))$. As one can observe, $\tilde{a}(t) \equiv a_S(t)$ in a long trait. The simulation was done using AMPL-Ipopt on the server NEOS, and pictures have been made with Python

The surprising fact given by Theorem 3.7 is that the minimizing strategy is not only of bang–bang type but may also assume some values along a *singular arc*, given by $a_S(t)$. This possibility is realized in some concrete cases, as we verified by running some numerical simulations, whose results can be visualized in Fig. 3.2.

3.5 Discussion of the Results

We now give the interpretation of the analyses of the aggressive competition model that we propose. We discuss the results in the key of the economic model because it seems to us the interpretation most applicable to reality, while we again dissociate ourselves from the use of violence between human beings.

We emphasize that, unlike the Lotka–Volterra model, our system allows for the case in which one of the two firms (the "attacked" one) fails in finite time (cf. Theorem 3.2). Once the competition is eliminated, the first firm obtains a monopoly in the market, with all the benefits of the case and the consequent problems for consumers.

In the absence of aggression ($a = 0$), the two companies would eventually end up on a coexisting equilibrium in the market (see Proposition 6.3). When bringing in an aggressive strategy, we can generally see three aspects. When the aggressive company (u) is already clearly preponderant in the market compared to the competitor (v), regardless of the strategy adopted, it will succeed in wiping out the competitor from the market. In contrast, if the competitor is very established in the market, it will not always be possible to supplant it, even in the case where the aggressive firm is more efficient than the opponent ($\rho < 1$), and in this case the aggressive strategy harms the attacking firm much more.

Interestingly, in cases close to the limit, the aggressive firm could only completely supplant the second with aggressive strategies that would bring itself close to failure.

As for the cases of intermediate initial situations, we observe that indeed the choice of strategy influences the final outcome. We can draw the following conclusions, depending on whether the market is initially saturated or not and whether the aggressive firm is more or less efficient than its competitor in reusing its capital to generate more.

We emphasize that the case of an oversaturated market may occur, for example, when there is a shrinking pool of buyers due to an economic or demographic crisis or when some consumers have the products of both companies for a period out of curiosity.

A more specific description of the different scenarios can be summarized as follows:

- *Scenario 1: The two firms have the same efficiency ($\rho = 1$).*
  In this case, the choice of the strategy does not affect the outcome. Namely, one of the two firms will eventually prevail according only to the initial conditions, independently of the aggressive strategy. The strategy just modulates the speed of the dynamics. This is shown in Theorem 3.3 part (i) and Lemma 7.2.
- *Scenario 2: The first firm is more efficient than the second ($\rho < 1$).*
  If the market is not oversaturated ($u+v \leqslant 1$), then it is more convenient for the first firm to "let the market flow" and use light aggression (a very small). If the market, on the other hand, is oversaturated ($u+v > 1$), it is convenient for the first firm to adopt a very aggressive strategy (a very large), to bring the market to an undersaturated condition which is, however, as convenient as possible for it, and once this intermediate goal is achieved, to continue with less intense aggression. In particular, in this latter case,

nonconstant strategies are better than constant strategies. These results are contained in Theorems 3.3, 3.4, and 3.5 and Proposition 7.1, part 1.

- *Scenario 3: The first firm is less efficient than the second ($\rho > 1$).*
  If the market is not oversaturated ($u + v \leqslant 1$), a very aggressive strategy allows the first firm to obtain monopoly even in cases where light aggressiveness would not allow it.
  If, on the contrary, the market is oversaturated ($u + v < 1$), it is convenient for the first firm not to use an aggressive strategy ($a = 0$) until the market reaches an unsaturated state; then, adopting a strongly aggressive strategy, the firm will be able to eventually wipe the rival out (as long as the initial data belong to the set presented in the formula (7.3)).
  These results are presented in Theorems 3.3, 3.4, and 3.5 and Proposition 7.1, part 2.

We also analyzed, in the case of initial situations that allow the first firm for "winning" strategies, which strategy eliminates competitors from the market in the fastest way possible. What is highlighted is that this strategy can be very sophisticated; in particular, it can alternate between very high and very low values of aggressiveness or follow a certain function (the singular arc function a_s defined in Theorem 3.7). Although it is difficult to give a general expression for the fastest strategy, it is possible to calculate or simulate it numerically from the initial data using well-known optimal control tools (see Fig. 3.2 and the proof of Theorem 3.7 in Sect. 7.6).

Toolbox 4

We start with some technical notation that we will often use in this work, in addition to the basic ones presented in Sect. 3.1. We will sometimes indicate the solutions $(u(t), v(t))$ of this system by $\phi_p(t)$, where $p = (u(0), v(0))$, in order to stress out the dependence on the initial position.

Throughout this monograph, solutions, trajectories, and orbits are always associated with system (2.1).

Given $(u_0, v_0) \in [0, 1] \times [0, 1]$ such that $T_s(u_0, v_0) = +\infty$, we define the ω-*limit set* of (u_0, v_0) as

$$\omega(u_0, v_0) := \big\{ (x, y) \in \mathbb{R}^2 \text{ s.t.}$$
$$\phi_{(u_0, v_0)}(t) \in [0, 1] \times [0, 1] \text{ for all } t \geqslant 0,$$
$$\text{and there exists } \{t_i\}_{i \in \mathbb{N}} \text{ s.t. } t_i \to +\infty$$
$$\text{and } \lim_{i \to +\infty} \phi_{(u_0, v_0)}(t_i) = (x, y) \big\}.$$

We also define the limit in the past as the α-*limit set* of (u_0, v_0) if $\phi_{(u_0, v_0)}(t) \in [0, 1] \times [0, 1]$ for all $t \leqslant 0$, that is,

$$\alpha(u_0, v_0) := \big\{ (x, y) \in \mathbb{R}^2 \text{ s.t.}$$
$$\phi_{(u_0, v_0)}(t) \in [0, 1] \times [0, 1] \text{ for all } t \leqslant 0,$$
$$\text{and there exists } \{t_i\}_{i \in \mathbb{N}} \text{ s.t. } t_i \to -\infty$$
$$\text{and } \lim_{i \to +\infty} \phi_{(u_0, v_0)}(t_i) = (x, y) \big\}.$$

We will refer to a periodic trajectory as a *closed orbit* (see, e.g., [77]).

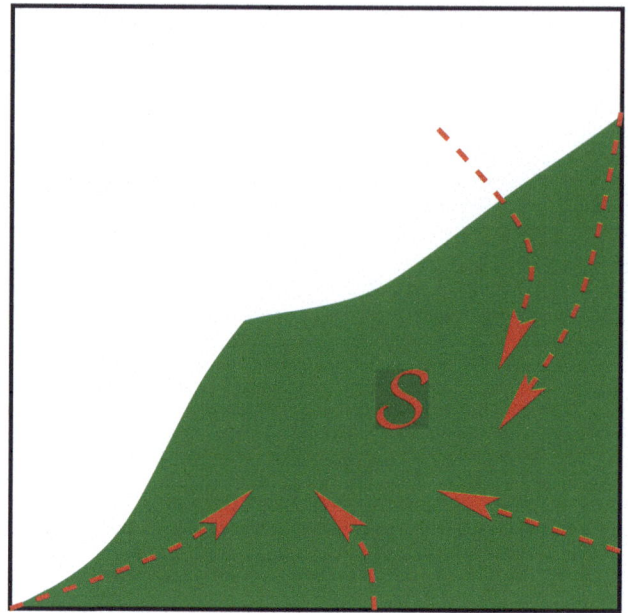

Fig. 4.1 Trajectories entering $\mathcal{S}$

Also, when we talk about *open* or *closed* sets contained in $[0, 1] \times [0, 1]$, it is always understood with respect to the relative topology of $[0, 1] \times [0, 1]$.

Definition 4.1 Given an open set $\mathcal{S}$ of $[0, 1] \times [0, 1]$, we say that a trajectory *enters* $\mathcal{S}$ through a point $p \in \partial \mathcal{S}$ if, letting $(u(t), v(t))$ be a solution generating such a trajectory, there exist a time $T \geqslant 0$ and a strictly decreasing sequence $(t_n)_{n \in \mathbb{N}}$ converging to T such that:

- $(u(T), v(T)) = p$.
- $(u(t_n), v(t_n)) \in \mathcal{S}$ for all $n \in \mathbb{N}$.

See Fig. 4.1 for a sketch of this notion of entering.

Remark 4.2 Notice that the side $\{0\} \times [0, 1]$ coincides with the orbit starting in the equilibrium $(0, 0)$ and arriving to the equilibrium $(0, 1)$. We mostly consider sets that are subgraphs of a continuous function (see Lemma 4.6). Thus, by the uniqueness of the Cauchy problem, no trajectory can enter these sets through the left side $\{0\} \times [0, 1]$.

Definition 4.3 If $\mathcal{S}$ is a closed set in the topology of $[0, 1] \times [0, 1]$, we also say that a trajectory *exits* the set $\mathcal{S}$ through a point $p \in \partial \mathcal{S}$ if it enters $[0, 1] \times [0, 1] \setminus \mathcal{S}$ through p (see Fig. 4.2).

4 Toolbox

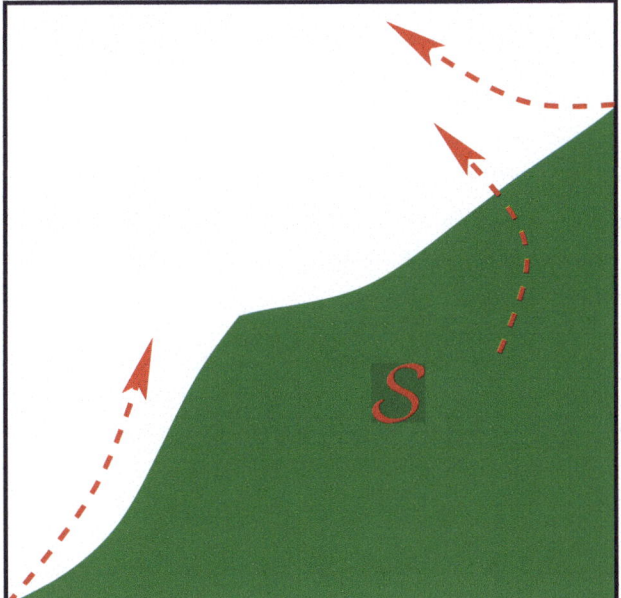

Fig. 4.2 Trajectories exiting S

Remark 4.4 As we already discussed in Remark 3.1, a trajectory either stays in $[0, 1] \times [0, 1]$ for all positive times or has a finite stopping time. This is why we do not take into account the situations in which a trajectory could exit "through the sides of $[0, 1] \times [0, 1]$" in all the rest of the monograph without further mention.

Now, we enunciate a useful lemma about entrance and exit of the trajectories in a set. In the following statement, we have a set S that is the set of points given by the subgraph of a function $g(u)$.

We define the outward unit normal vector to the surface ∂S at the point $(\check{u}, g(\check{u}))$ by

$$\nu = \left(-\frac{g'(\check{u})}{\sqrt{1 + (g'(\check{u}))^2}}, \frac{1}{\sqrt{1 + (g'(\check{u}))^2}}\right) \quad (4.1)$$

whenever $g'(\check{u})$ exists. We observe that "outward" here is intended with respect to the subgraph of g.

We also extend this notion of outward unit normal vector at points where ∂S has a corner or a vertical tangency[1] (hence $g'(\check{u})$ does not exist or is infinite). In this situation,

---

[1] Actually, later on, we will focus our attention on monotone increasing functions g; thus the case in which $\lim_{u \to \check{u}} g'(u) = -\infty$ will not be used (we mentioned it at this level just for completeness).

- If $\lim_{u \to \check{u}} g'(u) = +\infty$, then we take $\nu = (-1, 0)$ (i.e., the one obtained from (4.1) by formally replacing $g'(\check{u})$ by $+\infty$).
- If $\lim_{u \to \check{u}} g'(u) = -\infty$, then we take $\nu = (1, 0)$ (i.e., the one obtained from (4.1) by formally replacing $g'(\check{u})$ by $-\infty$).
- If $\ell_+ = \lim_{u \to \check{u}^+} g'(u)$ and $\ell_- = \lim_{u \to \check{u}^-} g'(u)$ exist (possibly infinite) but are different, we admit that $\partial \mathcal{S}$ has two outward unit normal vectors at $\check{u}$ (i.e., the ones obtained from (4.1) by replacing $g'(\check{u})$ by ℓ_- and ℓ_+, possibly using the conventions in the first two points on this list).

Furthermore, for the sake of clarity, for all $(\check{u}, \check{v}) \in \partial \mathcal{S}$ we denote by $N(\check{u}, \check{v})$ *the set of outward unit normal vectors to $\mathcal{S}$ at $(\check{u}, \check{v})$*. Notice that for the points $(\check{u}, g(\check{u}))$ this set has one element when g is differentiable at $\check{u}$ or one of the first two cases in the previous list occurs, while when the last case of the previous list occurs, it has two elements.

In this chapter we call

$$F(u, v) = u(1 - u - v - ac), \qquad G(u, v) = \rho v(1 - u - v) - ac,$$

so that system (2.1) becomes

$$\begin{cases} \dot{u} = F(u, v), & \text{for } t > 0, \\ \dot{v} = G(u, v), & \text{for } t > 0, \end{cases}$$

where F and G are locally Lipschitz-continuous functions.

In the forthcoming Lemma 4.6, we will require that for all $(\check{u}, g(\check{u})) \in \partial \mathcal{S}$, the trajectory starting at $(\check{u}, g(\check{u}))$ satisfies

$$(F(\check{u}, g(\check{u})), G(\check{u}, g(\check{u}))) \cdot \nu \geq 0 \qquad \text{for all } \nu \in N(\check{u}, g(\check{u})).$$

Remark 4.5 Notice that if $(\bar{u}, \bar{v})$ is not an equilibrium for the system, the trajectory starting at some $(\bar{u}, \bar{v})$ has the vector $(F(\check{u}, g(\check{u})), G(\check{u}, g(\check{u})))$ as tangent vector at $(\bar{u}, \bar{v})$. Hence, the scalar product $(F(\check{u}, g(\check{u})), G(\check{u}, g(\check{u}))) \cdot \nu$ gives us information on the relative position of the trajectory and the vector ν. In particular, if the scalar product has a sign, this tells us in which direction the trajectory crosses the graph of g.

However, we want to specify the structure of the set where the scalar product is equal to zero for at least one normal vector. In fact, we only treat the cases where the product is zero in a finite number of closed intervals and isolated singletons. The reason for this is to avoid pathological cases, i.e., when there is a dense sequence of singletons where the scalar product is zero.

4 Toolbox

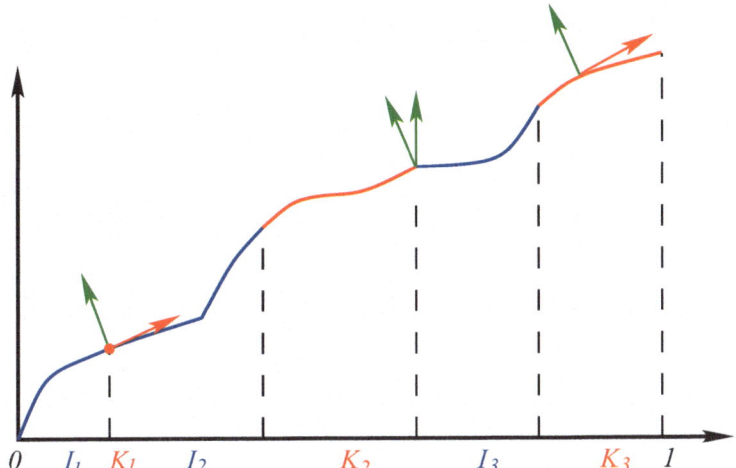

Fig. 4.3 A possible choice for the function $g(u)$ and intervals $\{I_k\}_{k \in A}$, $\{K_j\}_{j \in C}$ satisfying the hypothesis of Lemma 4.6. Plotted in blue are the traits for $u \in I_k$ with $k = 1, 2, 3$. Plotted in red are the points and traits for $u \in K_j$ with $j = 1, 2, 3$. Plotted in green are the outward unit normal vectors and in red the tangent vectors to the graph of g

Let C be a finite subset of $\mathbb{N}$, and let $\{K_j\}_{j \in C}$ be a collection of closed intervals, possibly coinciding with singletons. Then, notice that in the topology of $[0, 1]$, the set

$$[0, 1] \setminus \bigcup_{j \in C} K_j$$

is an open set and can be written as a collection of open intervals $\{I_k\}_{k \in A}$ for some finite subset A of $\mathbb{N}$. This setting can be visualized in Fig. 4.3.

Now we are ready to give the following:

Lemma 4.6 *Let*

$$\mathcal{S} := \{(u, v) \in [0, 1] \times [0, 1] \ s.t. \ v < g(u)\},$$

where $g : [0, 1] \to [0, +\infty)$ is a monotone increasing, continuous function such that there exists a finite (possibly empty) set $Z \subset [0, 1]$ for which $g \in C^1([0, 1] \setminus Z)$, and in addition the limits

$$\lim_{u \to z^{\pm}} g'(u)$$

exist for all $z \in Z$ (possibly distinct and possibly equal to $+\infty$). Let $N(\check{u}, g(\check{u}))$ be the set of outward unit normal vectors to $\mathcal{S}$ at $(\check{u}, g(\check{u}))$, as defined above. Let $\{I_k\}_{k \in A}$ and

$\{K_j\}_{j \in C}$ *be two finite collections of disjoint intervals in* $[0, 1]$, *the* I_k *being open and the* K_j *being closed (possibly coinciding with singletons) in the topology induced from* $[0, 1]$, *such that*

$$\bigcup_{k \in A} I_k \cup \bigcup_{j \in C} K_j = [0, 1].$$

Suppose that for all $\check{u} \in \bigcup_{k \in A} I_k$ *it holds that*

$$\min_{v \in N(\check{u}, g(\check{u}))} (F(\check{u}, g(\check{u})), G(\check{u}, g(\check{u}))) \cdot v > 0 \qquad (4.2)$$

and that for all $\check{u} \in \bigcup_{j \in C} K_j$ *it holds that*

$$\min_{v \in N(\check{u}, g(\check{u}))} (F(\check{u}, g(\check{u})), G(\check{u}, g(\check{u}))) \cdot v = 0. \qquad (4.3)$$

Then, no trajectory enters $\mathcal{S}$ *through a point of the set*

$$\partial \mathcal{S} \cap \{(u, v) \ \text{s.t.} \ u \in [0, 1], \ v = g(u)\}.$$

Proof Let us call

$$\mathcal{G} := \{(u, v) \ \text{s.t.} \ u \in [0, 1], \ v = g(u)\}.$$

We suppose that there exists $u_M \in (0, 1]$ such that $g(u) < 1$ for all $u \in [0, u_M)$ and $(u_M, g(u_M)) \in \partial([0, 1] \times [0, 1])$; the case $g(u) > 1$ for all u is trivial because in this case $\mathcal{S}$ coincides with $[0, 1] \times [0, 1]$ and $\partial \mathcal{S} \cap \mathcal{G} = \emptyset$.

We argue by contradiction and suppose that there exists a trajectory entering $\mathcal{S}$ through a point $(\check{u}, g(\check{u}))$, with $\check{u} \in [0, u_M]$. Let $(\widehat{u}, \widehat{v})$ be the solution generating such a trajectory, with (up to translation in time) $(\widehat{u}(0), \widehat{v}(0)) = (\check{u}, g(\check{u}))$.

Notice that either $\check{u} \in I_k$ for a unique $k \in A$ or $\check{u} \in K_j$ for a unique $j \in C$.

The assumptions on the regularity of $g(u)$ imply that $\partial \mathcal{S} \cap \mathcal{G}$ has a unique normal vector, except at the corner points, where it has 2.

We now distinguish three cases. In the first two cases, we distinguish whether g is differentiable at $\check{u}$ or g is not differentiable at $\check{u}$. In the last case we analyze the role of the extrema $\check{u} = 0$ and $\check{u} = u_M$.

Case 1: $\check{u} \notin Z \cup \{0, u_M\}$.

In this case, g is differentiable in $\check{u}$. Notice that since g is piecewise C^1 with a finite number of non-differentiability points, then there exists a neighborhood of $\check{u}$ (in $[0, 1]$) where g is C^1.

If $\check{u} \in I_k$ for some $k \in A$, then by (4.2) and Remark 4.5 one has a contradiction with the fact that $(\widehat{u}, \widehat{v})$ enters $\mathcal{S}$ through the point $(\check{u}, g(\check{u}))$.

If $\check{u} \in \mathrm{int}\, K_j$ for some $j \in C$, then by (4.1) and (4.3) there exists a neighborhood U of $\check{u}$ such that for each $u \in U$ it holds that

$$-g'(u)F(u, g(u)) + G(u, g(u)) = 0. \tag{4.4}$$

Let φ be the solution of the Cauchy problem

$$\begin{cases} \varphi'(t) = F(\varphi(t), g(\varphi(t))), \\ \varphi(0) = \check{u}, \end{cases} \tag{4.5}$$

which exists for $t \geq 0$ sufficiently small. It follows from (4.4) that $(\varphi(t), g(\varphi(t)))$ solves (2.1) for $t \geq 0$ sufficiently small; hence for these t's, we have that

$$(\widehat{u}(t), \widehat{v}(t)) = (\varphi(t), g(\varphi(t))).$$

This contradicts the fact that the associated trajectory enters $\mathcal{S}$ through $(\check{u}, g(\check{u}))$.

If $\check{u} \in \partial K_j$ for some fixed $j \in C$, we distinguish two cases depending on whether K_j is a singleton or not.

If $K_j = \{\check{u}\}$, recalling that the trajectory generated by $(\hat{u}(t), \hat{v}(t))$ enters $\mathcal{S}$ through $(\check{u}, g(\check{u}))$ (at time 0), there exists an arbitrarily small time $\tau > 0$ such that $(\hat{u}(\tau), \hat{v}(\tau)) \in \mathcal{S}$. Then, by the continuous dependence with respect to initial data and because $\mathcal{S}$ is an open set, there exists a ball B centered at $(\check{u}, g(\check{u}))$ and of radius ε sufficiently small such that $\phi_q(\tau) \in \mathcal{S}$ for every $q \in B$. In addition, for given $\delta > 0$, up to reducing the time $\tau > 0$ and the radius ε of the ball B if need be, we have that

$$\forall t \in [0, \tau], \ \forall q \in B, \quad \phi_q(t) \in (\check{u} - \delta, \check{u} + \delta) \times [0, 1].$$

As a consequence, for $q \in B \setminus \mathcal{S}$, there must exist $\bar{t} \in (0, \tau)$ and $\underline{u} \in (\check{u} - \delta, \check{u} + \delta)$ such that

$$\phi_q(\bar{t}) = (\underline{u}, g(\underline{u})) \in \partial \mathcal{S}.$$

Then, taking δ smaller than the distance between the point $\check{u}$ and the compact set $\bigcup_{j' \neq j} K_{j'}$, we infer

$$\underline{u} \in K_j \cup \bigcup_{k \in A} I_k.$$

In addition, since g is differentiable at $\check{u}$, it must be differentiable in a neighborhood of $\check{u}$, by hypothesis; hence, for even smaller δ, we have that g is differentiable at $\underline{u}$. But we have shown before that trajectories cannot enter $\mathcal{S}$ through a point $(u, g(u))$ with $u \in I_k$ and g

differentiable at u, and therefore it has to be $\underline{u} = \check{u}$. Summing up, we have shown that

$$\forall q \in B \setminus \mathcal{S}, \; \exists \bar{t} \in (0, \tau), \quad \phi_q(\bar{t}) = (\check{u}, g(\check{u})). \tag{4.6}$$

This means that $B \setminus \mathcal{S}$ is a subset of the trajectory

$$\{\phi_{(\check{u}, g(\check{u}))}(t) \; : \; t \in [-\tau, 0]\},$$

which is impossible because the trajectory has zero measure, whereas $B \setminus \mathcal{S}$ has positive measure, being g a continuous function.

Thus, we are left with the case where $\check{u} \in \partial K_j$ and K_j is an interval. Then, it must be that $\check{u} \in \partial I_k$ for some $j \in A$. Without loss of generality, we can suppose that $I_k = (u_1, \check{u})$ and $K_j = [\check{u}, u_2]$ for some $0 \leqslant u_1 < \check{u} < u_2 \leqslant u_M$.

Let us denote by μ the vector that is tangent to the graph of $g(u)$ at $(\check{u}, g(\check{u}))$ and such that μ has positive components.

We now distinguish three cases.

If $(F(\check{u}, g(\check{u})), G(\check{u}, g(\check{u}))) \cdot \mu = 0$, then this and (4.3) give that $(\check{u}, g(\check{u}))$ is an equilibrium; hence no trajectory can enter through $(\check{u}, g(\check{u}))$.

If $(F(\check{u}, g(\check{u})), G(\check{u}, g(\check{u}))) \cdot \mu > 0$, then at least one between $F(\check{u}, g(\check{u}))$ and $G(\check{u}, g(\check{u}))$ must be positive. Moreover, by (4.3), we have

$$-g'(\check{u}) F(\check{u}, g(\check{u})) + G(\check{u}, g(\check{u})) = 0. \tag{4.7}$$

Notice that if $F(\check{u}, g(\check{u})) < 0$, by the fact that g is increasing and (4.7), we get that $G(\check{u}, g(\check{u})) < 0$, which contradicts our assumption.

Consequently,

$$F(\check{u}, g(\check{u})) > 0. \tag{4.8}$$

Also, there is a right neighborhood U of $\check{u}$ such that for all $u \in U$ (4.4) holds true. Let $\varphi(t)$ be a solution of (4.5). Then, by (4.8), we get that $\varphi(t)$ is increasing; thus $\varphi(t) \in U$ for small t. Hence, (4.4) holds true, and $(\varphi(t), g(\varphi(t)))$ is a solution of the system in (2.1) starting at $(\check{u}, g(\check{u}))$ and laying on the graph of g for small t. This contradicts the fact that a trajectory enters $\mathcal{S}$ through $(\check{u}, g(\check{u}))$.

We are left with the case $(F(\check{u}, g(\check{u})), G(\check{u}, g(\check{u}))) \cdot \mu < 0$. Then, arguing as in the previous case, we can prove that $F(\check{u}, g(\check{u})) < 0$. By continuity, we have that

$$F(u, g(u)) < 0 \tag{4.9}$$

for $u \in U$, being U a right neighborhood of $\check{u}$.

Let $\bar{u} \in U$. Then, by arguing as in the previous case, we have that the trajectory of $(\bar{u}, g(\bar{u}))$ can be written as $(\psi(t), g(\psi(t)))$, where $\psi(t)$ is a solution of

4 Toolbox

$$\begin{cases} \psi'(t) = F(\psi(t), g(\psi(t))), \\ \psi(0) = \bar{u}. \end{cases}$$

By (4.9), we have that $\psi(t)$ is decreasing. Moreover, since (4.9) holds true for all $u \in U$, it follows that $\psi(\bar{t}) = \check{u}$ for some $\bar{t} > 0$. Hence, since for all $u \in U \setminus \{\check{u}\}$ the point $(u, g(u))$ belongs to the trajectory of $(\bar{u}, g(\bar{u}))$, no trajectory can enter $\mathcal{S}$ through $(u, g(u))$ for $u \in U \setminus \{\check{u}\}$.

Now take $V \subset I_k$, where V is a left neighborhood of $\check{u}$ where g is differentiable. Then for all $u \in V \setminus \{\check{u}\}$, as seen in the first case, no trajectory can enter $\mathcal{S}$ through $(u, g(u))$.

Now, take a ball B of radius ε centered at $(\check{u}, g(\check{u}))$. By arguing as in the case when K_j is a singleton, we can prove that for all $q \in B \setminus \mathcal{S}$ (which has positive measure), it holds that $\phi_q(\bar{t}) = (\check{u}, g(\check{u}))$ for some $\bar{t} \in (0, \delta)$ because no trajectory can enter through other points in a neighborhood of $\check{u}$. So

$$\phi_{B \setminus \mathcal{S}}(t) \subset \Upsilon \quad \text{for } t \in (\delta, 2\delta),$$

where

$$\Upsilon := \big\{(u, v) \in [0, 1] \times [0, 1] \, ; \, \text{s.t.} \\ (u, v) = \phi_{(\check{u}, g(\check{u}))}(t) \text{ for } t \in [0, 2\delta]\big\}.$$

But Υ has measure 0, and $\phi_{B \setminus \mathcal{S}}(t)$ has positive measure, thus giving a contradiction.

Case 2: $\check{u} \in Z \setminus \{0, u_M\}$.

Since $\check{u} \in Z$, then g is not differentiable at $\check{u}$. By hypothesis, there are a finite number of non-differentiability points; hence they are isolated points. So, given a neighborhood U of $\check{u}$, it holds that g is differentiable in $U \setminus \{\check{u}\}$. Hence, no trajectory can enter $\mathcal{S}$ through a point of the form $(u, g(u))$ with $u \in U \setminus \{\check{u}\}$, in light of Case 1.

Now, take a ball B of radius ε centered in $(\check{u}, g(\check{u}))$. By the continuity of g, we see that $B \setminus \mathcal{S}$ has positive measure. By arguing as in the case when K_j is a singleton, for all $q \in B \setminus \mathcal{S}$ it holds that $\phi_q(\bar{t}) = (\check{u}, g(\check{u}))$ for some $\bar{t} \in (0, \delta)$ because no trajectory can enter through other points in a neighborhood of $\check{u}$. So

$$\phi_{B \setminus \mathcal{S}}(t) \subset \Upsilon \quad \text{for } t \in (\delta, 2\delta),$$

where

$$\Upsilon := \big\{(u, v) \in [0, 1] \times [0, 1] \text{ s.t.} \\ (u, v) = \phi_{(\check{u}, g(\check{u}))}(t) \text{ for } t \in [0, 2\delta]\big\}.$$

But Υ has measure 0, and $\phi_{B\setminus S}(t)$ has positive measure, thus providing the desired contradiction.

Case 3: $\check{u} = 0$ or $\check{u} = u_M$.

For the sake of concreteness, let us consider the case $\check{u} = 0$, the other one being analogous. Let us consider a right neighborhood U of $\check{u} = 0$ (in the topology of $[0, 1]$). By Cases 1 and 2,

$$\text{no trajectory can enter } S \text{ through a point} \atop \text{of the form } (u, g(u)) \text{ with } u \in U \setminus \{0\}. \qquad (4.10)$$

Suppose that a trajectory enters S through $(0, g(0))$. Also, consider a ball B centered at $(0, g(0))$ of radius ε sufficiently small and the set

$$D = B \cap ([0, 1] \times [0, 1]) \setminus S,$$

which has positive measure.

By continuity with respect to initial data, there exists $\delta > 0$ such that $\phi_q(\delta) \in S$ for every $q \in D$.

As a consequence, there exists a point $(\bar{u}, \bar{v}) \in \partial S$ such that $\phi_q(\bar{t}) = (\bar{u}, \bar{v})$ for some $\bar{t} \in (0, \delta)$. Notice that $(\bar{u}, \bar{v})$ cannot be of the form $(u, g(u))$ with $u \in U \setminus \{\check{u}\}$, in light of (4.10).

Also, the trajectory of q is contained in $[0, 1] \times [0, 1]$ for all $t < T_s(q)$ (where T_s is the stopping time defined in (3.1)). Thus, it cannot be that $\phi_q(\bar{t}) = (0, \underline{v})$ with $\underline{v} \leqslant g(0)$, unless $\phi_q(\bar{\tau}) = (u_0, g(\check{u}))$ for some $\bar{\tau} \in (0, \delta)$.

Therefore, for $t \in (\delta, 2\delta)$, we have that $\phi_D(t) \subset \Upsilon$, where

$$\Upsilon := \big\{(u, v) \in [0, 1] \times [0, 1] \text{ s.t.} \\ (u, v) = \phi_{(\check{u}, g(\check{u}))}(t) \text{ for } t \in [0, 2\delta]\big\}.$$

But Υ has measure 0, and $\phi_{B\setminus S}(t)$ has positive measure, thus providing the desired contradiction and completing the proof of Lemma 4.6. □

We also provide a stronger statement for exiting trajectories. Notice that here we take a closed set S to use Definition 4.3 of exiting trajectories.

Lemma 4.7 *Let*

$$S := \{(u, v) \in [0, 1] \times [0, 1] \text{ s.t. } v \leqslant g(u)\},$$

where g satisfies the same hypotheses as in Lemma 4.6. Suppose also that I_k and K_j are as in Lemma 4.6.

4 Toolbox

Suppose that for all $\check{u} \in \bigcup_{k \in A} I_k$ it holds that

$$\max_{v \in N(\check{u}, g(\check{u}))} (F(\check{u}, g(\check{u})), G(\check{u}, g(\check{u}))) \cdot v < 0 \tag{4.11}$$

and that for all $\check{u} \in \bigcup_{j \in C} K_j$ it holds that

$$\max_{v \in N(\check{u}, g(\check{u}))} (F(\check{u}, g(\check{u})), G(\check{u}, g(\check{u}))) \cdot v = 0. \tag{4.12}$$

Then, no trajectory exits $\mathcal{S}$.

Proof Let us suppose by contradiction that there exists a trajectory exiting $\mathcal{S}$ through a point $(\check{u}, \check{v})$.

If

$$(\check{u}, \check{v}) \in \partial \mathcal{S} \cap \{(u, v) \text{ s.t. } u \in [0, 1], v = g(u)\},$$

then repeating the arguments of Lemma 4.6 we get a contradiction.

We also observe that no trajectory can exit $\mathcal{S}$ by leaving $[0, 1] \times [0, 1]$. This rules out all the possible cases. □

Finally, to lighten the text, all along this monograph, we will call *outward normal derivative* at some point $(\check{u}, \check{u}) \in \partial \mathcal{S}$ the scalar product

$$(F(\check{u}, \check{v}), G(\check{u}, \check{v})) \cdot v$$

with $v \in N(\check{u}, \check{v})$.

Also, we call the *inward normal derivative* at some point $(\check{u}, \check{v}) \in \partial \mathcal{S}$ the scalar product

$$-(F(\check{u}, \check{v}), G(\check{u}, \check{v})) \cdot v$$

with $v \in N(\check{u}, \check{v})$.

Basins of Attractions

In this chapter we provide some useful results on the behavior of the solutions of the system in (2.1) and on the basins of attraction in the case of constant strategies a. In particular, we provide the proof of Theorem 3.2, and we state a characterization of the sets $\mathcal{B}$ and $\mathcal{E}$ given in (3.3) and (3.4), respectively, see Propositions 5.9.

This material will be extremely useful for the analysis of the strategy that we operate later.

We are now in a position to derive the first three statements of Theorem 3.2.

Proof of (i), (ii), and (iii) of Theorem 3.2 We first consider equilibria with the first coordinate $u = 0$. In this case, from the second equation in (2.1), we have that the equilibria must satisfy $\rho v(1 - v) = 0$; thus $v = 0$ or $v = 1$. As a consequence, $(0, 0)$ and $(0, 1)$ are two equilibria of the system.

Next, we consider equilibria with the first coordinate $u > 0$. From the first equation in (2.1), we get

$$1 - u - v - ac = 0, \tag{5.1}$$

while, from the second one,

$$\rho v(1 - u - v) - au = 0. \tag{5.2}$$

Putting together (5.1) and (5.2), we get (3.5).

From now on, we distinguish the three situations in (i), (ii), and (iii) of Theorem 3.2.

(i) If $0 < ac < 1$, we have that the point (u_s, v_s) given in (3.5) lies in $(0, 1) \times (0, 1)$. As a result, in this case the system has three equilibria, given by $(0, 0)$, $(0, 1)$, and (u_s, v_s).

The Jacobian of the system (2.1) is

$$J(u, v) = \begin{pmatrix} 1 - 2u - v - ac & -u \\ -\rho v - a & \rho(1 - u - 2v) \end{pmatrix}. \tag{5.3}$$

At the point $(0, 0)$, the matrix has eigenvalues $\rho > 0$ and $1 - ac > 0$; thus $(0, 0)$ is a source.

At the point $(0, 1)$, the Jacobian (5.3) has eigenvalues $-ac < 0$ and $-\rho < 0$; thus $(0, 1)$ is a sink.

At the point (u_s, v_s), by exploiting the relations (5.1) and (5.2) we have that

$$J(u_s, v_s) = \begin{pmatrix} -u_s & -u_s \\ -\rho v_s - a & \rho(ac - v_s) \end{pmatrix},$$

which, by the change of basis given by the matrix,

$$\begin{pmatrix} -\frac{1}{u_s} & 0 \\ -\frac{1}{u_s}\left[\left(\frac{u_s}{c} + a\right)\left(\frac{\rho c - c}{1+\rho c}\right) + ac\right] & \frac{\rho c - c}{1+\rho c} \end{pmatrix},$$

becomes

$$\begin{pmatrix} 1 & 1 \\ ac & \rho ac \end{pmatrix}.$$

The characteristic polynomial of this matrix is

$$\lambda^2 - \lambda(1 + \rho ac) + \rho ac - ac,$$

which has two real roots, as one can see by inspection. Hence, the matrix $J(u_s, v_s)$ has two real eigenvalues.

Moreover, the determinant of $J(u_s, v_s)$ is

$$-\rho acu_s - au_s < 0,$$

which implies that $J(u_s, v_s)$ has one positive and one negative eigenvalues. These considerations give that (u_s, v_s) is a saddle point. This completes the proof of (i) in Theorem 3.2.

(ii) and (iii) We assume that $ac \geqslant 1$. We observe that the equilibrium described by the coordinates (u_s, v_s) in (3.5) coincides with $(0, 0)$ for $ac = 1$ and lies outside $[0, 1] \times [0, 1]$ for $ac > 1$. As a result, when $ac \geqslant 1$, the system has two equilibria, given by $(0, 0)$ and $(0, 1)$.

Looking at the Jacobian in (5.3), one sees that at the point $(0, 1)$, it has eigenvalues $-ac < 0$ and $-\rho < 0$, and therefore $(0, 1)$ is a sink when $ac \geqslant 1$.

Furthermore, from (5.3) one finds that if $ac > 1$, then $J(0, 0)$ has the positive eigenvalue ρ and the negative eigenvalue $1 - ac$; thus $(0, 0)$ is a saddle point.

If instead $ac = 1$, then $J(0, 0)$ has one positive eigenvalue and one null eigenvalue, as desired. □

To complete the proof of Theorem 3.2, we will deal with the cases $ac \neq 1$ and $ac = 1$ separately. This analysis will be performed in the forthcoming Sects. 5.1 and 5.2, respectively. The completion of the proof of Theorem 3.2 will then be given in Sect. 5.3.

5.1 Characterization of $\mathcal{M}$ When $ac \neq 1$

We point out that in the proof of (i) and (ii) in Theorem 3.2 we found a saddle point in both cases. By the Stable Manifold Theorem (see, e.g., [77]), the point (u_s, v_s) in (3.5) in the case $0 < ac < 1$ and the point $(0, 0)$ in the case $ac > 1$ have a stable manifold and an unstable manifold. These manifolds are unique, they have dimension 1, and they are tangent to the eigenvectors of the linearized system.

We will denote by $\mathcal{M}$ the stable manifold associated with these saddle points. Since we are interested in the dynamics in the square $[0, 1] \times [0, 1]$, with a slight abuse of notation we will only consider the restriction of $\mathcal{M}$ in $[0, 1] \times [0, 1]$.

We now analyze some properties of $\mathcal{M}$:

Proposition 5.1 *For $ac \neq 1$, the set $\mathcal{M}$ can be written as the graph of a unique increasing C^2 function*

$$\gamma : [0, u_\mathcal{M}] \to [0, v_\mathcal{M}]$$

for some

$$(u_\mathcal{M}, v_\mathcal{M}) \in \big(\{1\} \times [0, 1]\big) \cup \big((0, 1] \times \{1\}\big),$$

such that $\gamma(0) = 0$, $\gamma(u_\mathcal{M}) = v_\mathcal{M}$ and

- *If $0 < ac < 1$, $\gamma(u_s) = v_s$, and in $u = u_s$, the function $\gamma(u)$ is tangent to the line $(v - v_s)c - (u - u_s) = 0$.*
- *If $ac > 1$, in $u = 0$, the function γ is tangent to the line $(\rho - 1 + ac)v - au = 0$.*

As a by-product of the proof of Proposition 5.1, we also obtain some useful information on the structure of the stable manifold and the basins of attraction, which we summarize here below:

Corollary 5.2 *Suppose that $0 < ac < 1$. Then, the curves (5.1) and (5.2), loci of the points such that $\dot{u} = 0$ and $\dot{v} = 0$, respectively, divide the square $[0, 1] \times [0, 1]$ into four regions:*

$$\begin{aligned} \mathcal{A}_1 &:= \{(u, v) \in [0, 1] \times [0, 1] \text{ s.t } \dot{u} \leqslant 0, \ \dot{v} \geqslant 0\}, \\ \mathcal{A}_2 &:= \{(u, v) \in [0, 1] \times [0, 1] \text{ s.t } \dot{u} \leqslant 0, \ \dot{v} \leqslant 0\}, \\ \mathcal{A}_3 &:= \{(u, v) \in [0, 1] \times [0, 1] \text{ s.t } \dot{u} \geqslant 0, \ \dot{v} \leqslant 0\}, \\ \text{and} \quad \mathcal{A}_4 &:= \{(u, v) \in [0, 1] \times [0, 1] \text{ s.t } \dot{u} \geqslant 0, \ \dot{v} \geqslant 0\}. \end{aligned} \quad (5.4)$$

Furthermore, the sets $\mathcal{A}_1 \cup \mathcal{A}_4$ and $\mathcal{A}_2 \cup \mathcal{A}_3$ are separated by the curve $\dot{v} = 0$, given by the graph of the continuous function

$$\sigma(v) := 1 - \frac{\rho v^2 + a}{\rho v + a}, \quad (5.5)$$

which satisfies $\sigma(0) = 0$, $\sigma(1) = 0$, and $0 < \sigma(v) < 1$ for all $v \in (0, 1)$.
In addition,

$$\mathcal{M} \setminus \{(u_s, v_s)\} \text{ is contained in } \mathcal{A}_2 \cup \mathcal{A}_4, \quad (5.6)$$

$$(\mathcal{A}_3 \setminus \{(0, 0), (u_s, v_s)\}) \subseteq \mathcal{E} \quad (5.7)$$

and

$$\mathcal{A}_1 \setminus \{(u_s, v_s)\} \subset \mathcal{B}, \quad (5.8)$$

where the notation in (3.3) and (3.4) has been utilized.

To visualize the statements in Corollary 5.2, one can see Fig. 5.1.

Corollary 5.3 *Suppose that $ac > 1$. Then, we have that $\dot{u} \leqslant 0$ in $[0, 1] \times [0, 1]$, and the curve (5.2) divides the square $[0, 1] \times [0, 1]$ into the regions*

$$\begin{aligned} \mathcal{A}_1 &:= \{(u, v) \in [0, 1] \times [0, 1] \text{ s.t. } \dot{u} \leqslant 0, \ \dot{v} \geqslant 0\} \\ \text{and} \quad \mathcal{A}_2 &:= \{(u, v) \in [0, 1] \times [0, 1] \text{ s.t. } \dot{u} \leqslant 0, \ \dot{v} \leqslant 0\}. \end{aligned} \quad (5.9)$$

5.1 Characterization of $\mathcal{M}$ When $ac \neq 1$

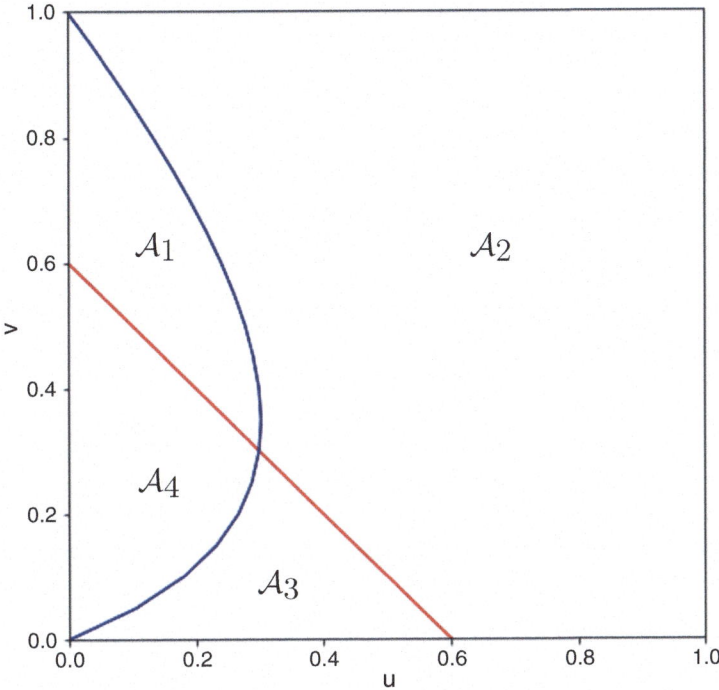

Fig. 5.1 Partition of $[0, 1] \times [0, 1]$ in the case $a = 0.8$, $c = 0.5$, $\rho = 2$, as given by (5.4). In red is given the curve $\dot{u} = 0$. In blue is given the curve $\dot{v} = 0$, parametrized by the function σ in (5.5)

Furthermore, the sets $\mathcal{A}_1$ and $\mathcal{A}_2$ are separated by the curve $\dot{v} = 0$, given by the graph of the continuous function σ given in (5.5).

In addition,

$$\mathcal{M} \subset \mathcal{A}_2. \tag{5.10}$$

Proposition 5.1 and Corollaries 5.2 and 5.3 are a bit technical but provide fundamental information to obtain a characterization of the sets $\mathcal{E}$ and $\mathcal{B}$, given in the forthcoming Proposition 5.9.

We now provide the proof of Proposition 5.1 (and, as a by-product, of Corollaries 5.2 and 5.3).

Proof of Proposition 5.1 and Corollaries 5.2 and 5.3 In the forthcoming argument, we treat separately the cases $0 < ac < 1$ and $ac > 1$. We start with the case $0 < ac < 1$ and divide the proof into three further parts.

Step 1: Localizing $\mathcal{M}$. Recall that an orbit is closed if the associate trajectory is periodic (see, e.g., [77]). We first observe that

$$\text{there are no closed orbits other than fixed points} \\ \text{entirely contained in } \mathcal{A}_i, i = 1, \ldots, 4, \tag{5.11}$$

because $\dot{u}$ and $\dot{v}$ have a sign.

With the notation introduced in (5.4), we prove that

$$\text{all points } (u_0, v_0) \in \mathcal{A}_3 \setminus \{(0, 0), (u_s, v_s)\} \\ \text{have } T_s(u_0, v_0) < +\infty. \tag{5.12}$$

To this aim, we show that

$$\text{no trajectory exits } \mathcal{A}_3. \tag{5.13}$$

First of all, we notice that no trajectory can exit $\mathcal{A}_3$ through $(0, 0)$ or (u_s, v_s) since they are equilibria.

We remark that the side connecting $(0, 0)$ and (u_s, v_s) can be written as the set of points belonging to

$$\{(u, v) \in [0, 1] \times (0, v_s) \text{ s.t. } u = \sigma(v)\},$$

where the function σ is defined in (5.5). In this set, it holds that $\dot{v} = 0$ and $\dot{u} > 0$, and thus the normal derivative pointing outward $\mathcal{A}_3$ is negative, so the trajectories cannot exit $\mathcal{A}_3$ passing through this side.

Furthermore, on the side connecting (u_s, v_s) with $(1 - ac, 0)$, which lies on the straight line $v = 1 - ac - u$, we have that $\dot{u} = 0$ and $\dot{v} < 0$ for $(u, v) \neq (u_s, v_s)$, so also here the outer normal derivative is negative. Therefore, the trajectories cannot exit $\mathcal{A}_3$ passing through this side either. This completes the proof of (5.13).

Now we show that

$$\text{there are no equilibria where a trajectory} \\ \text{lying in the interior of } \mathcal{A}_3 \text{ can converge.} \tag{5.14}$$

Indeed, if $p = (u_0, v_0) \in \mathcal{A}_3 \setminus \{(0, 0)\}$, then $u_0 > 0$. Now, $(0, 0) \notin \omega(p)$, because $\dot{u} \geqslant 0$ in $\mathcal{A}_3$. Also, for all $(u_0, v_0) \in \mathcal{A}_3 \setminus (u_s, v_s)$, we have that $v_0 < v_s$.

On the other hand, $\dot{v} \leqslant 0$ in $\mathcal{A}_3$, so no trajectory that is entirely contained in $\mathcal{A}_3$ can converge to (u_s, v_s). These observations prove (5.14).

As a consequence of (5.11), (5.13), and (5.14) and the Poincaré–Bendixson Theorem (see, e.g., [100], and see also Lemma 9.0.4), we have that all the points in the interior of $\mathcal{A}_3$ must have $T_s(u_0, v_0) < +\infty$.

These considerations complete the proof of (5.12). Accordingly, recalling the definition of $\mathcal{E}$ in (3.4), we see that

5.1 Characterization of $\mathcal{M}$ When $ac \neq 1$

$$(\mathcal{A}_3 \setminus \{(0,0), (u_s, v_s)\}) \subseteq \mathcal{E}. \tag{5.15}$$

In a similar way one can prove that all trajectories starting in $\mathcal{A}_1 \setminus \{(u_s, v_s)\}$ must converge to $(0, 1)$, which, recalling the definition of $\mathcal{B}$ in (3.3), implies that

$$(\mathcal{A}_1 \setminus \{(u_s, v_s)\}) \subset \mathcal{B}. \tag{5.16}$$

Thanks to (5.15) and (5.16), we have that the stable manifold $\mathcal{M}$ has no intersection with $\mathcal{A}_1 \setminus \{(u_s, v_s)\}$ and $\mathcal{A}_3 \setminus \{(0,0), (u_s, v_s)\}$, and therefore $\mathcal{M}$ must lie in $\mathcal{A}_2 \cup \mathcal{A}_4$.

Also, we know that $\mathcal{M}$ is tangent to an eigenvector in (u_s, v_s), and we observe that

$$(1, -1) \text{ is not an eigenvector of the linearized system.} \tag{5.17}$$

Indeed, if $(1, -1)$ were an eigenvector, then

$$\begin{pmatrix} 1 - ac - 2u_s - v_s & -u_s \\ -\rho v_s - a & \rho - \rho u_s - 2\rho v_s \end{pmatrix} \cdot \begin{pmatrix} 1 \\ -1 \end{pmatrix} = \lambda \begin{pmatrix} 1 \\ -1 \end{pmatrix},$$

so from the first component we would get $\lambda = 0$, which is not an eigenvalue of the saddle point (u_s, v_s). This establishes (5.17).

In light of (5.17), we conclude that $\mathcal{M} \setminus \{(u_s, v_s)\}$ must have intersection with both $\mathcal{A}_2$ and $\mathcal{A}_4$.

Step 2: Defining $\gamma(u)$ and tangential property. Since $\dot{u} > 0$ and $\dot{v} > 0$ in the interior of $\mathcal{A}_4$, the portion of $\mathcal{M}$ in $\mathcal{A}_4$ can be described globally as the graph of a monotone increasing smooth function $\gamma_1 : U \to [0, v_s]$, for a suitable interval $U \subseteq [0, u_s]$ with $u_s \in U$, and such that $\gamma_1(u_s) = v_s$.

We stress that, for $u > u_s$, the points $(u, v) \in \mathcal{M}$ belong to $\mathcal{A}_2$.

Similarly, in the interior of $\mathcal{A}_2$, we have that $\dot{u} < 0$ and $\dot{v} < 0$. Therefore, we find that $\mathcal{M}$ can be represented in $\mathcal{A}_2$ as the graph of a monotone increasing smooth function $\gamma_2 : V \to [v_s, 1]$, for a suitable interval $V \subseteq [u_s, 1]$ with $u_s \in V$, and such that $\gamma_2(u_s) = v_s$. Notice that in the second case the trajectories and the parametrization run in opposite directions.

Now, we define

$$\gamma(u) := \begin{cases} \gamma_1(u) & \text{if } u \in U, \\ \gamma_2(u) & \text{if } u \in V, \end{cases}$$

and we observe that it is an increasing smooth function locally parametrizing $\mathcal{M}$ around (u_s, v_s) (thanks to the Stable Manifold Theorem).

We point out that, in light of the Stable Manifold Theorem, the stable manifold $\mathcal{M}$ is globally parametrized by an increasing smooth function on a set $W \subset [0, 1]$.

We now study the tangent to $\gamma(u)$ at $u = u_s$. First, let us compute the derivative of $\gamma(u)$, which is given by

$$\frac{d\gamma(u)}{du} = \frac{dv(t)}{dt} \cdot \frac{dt}{du(t)} = \frac{\dot{v}}{\dot{u}}.$$

Notice that, as $u \to u_s^-$ and $v \to v_s^-$, owing the expression of u_s and v_s given in formula (3.5), we get

$$\lim_{\substack{u \to u_s \\ v \to v_s}} \frac{\dot{v}}{\dot{u}} = \lim_{\substack{u \to u_s \\ v \to v_s}} \frac{\rho v(1-u-v) - au}{u(1-u-v-ac)} = \frac{1}{c}.$$

From this we can say that

$$\gamma'(u) = \frac{1}{c}.$$

This gives us that $\gamma(u)$ is tangent to the line $(v - v_s)c - (u - u_s) = 0$, as desired.

Step 3: Showing that $\gamma(0) = 0$ and $\gamma(u_\mathcal{M}) = v_\mathcal{M}$ for some $(u_\mathcal{M}, v_\mathcal{M}) \in \partial([0,1] \times [0,1])$. We first prove that

$$\gamma(0) = 0. \tag{5.18}$$

For this, we claim that

$$\text{no trajectory enters } \mathcal{A}_4 \setminus \{(u_s, v_s)\}. \tag{5.19}$$

Indeed, it is easy to see that points in the form $(0, v_0)$ converge to $(0, 1)$. Hence, by the uniqueness of the trajectory passing through a nonfixed point, we get that no trajectory can enter through the side of $\mathcal{A}_4$ laying on $\{u = 0\}$.

No trajectory can enter through (u_s, v_s) or $(0, 0)$ since they are equilibria.

As for the side connecting $(0, 0)$ to (u_s, v_s), excluding the extrema, one has that $\dot{u} > 0$ and $\dot{v} = 0$, and so the inward pointing normal derivative is negative. Therefore, no trajectory can enter $\mathcal{A}_4$ on this side, see Remark 4.5.

Moreover, on the side connecting (u_s, v_s) to $(0, 1-ac)$, the inward pointing normal derivative is negative, because $\dot{u} = 0$ and $\dot{v} > 0$; thus we have that no trajectory can enter $\mathcal{A}_4$ on this side either. These considerations prove (5.19).

Furthermore, by 5.11, we have that no closed orbits are allowed in $\mathcal{A}_4$.

From (5.19), (5.11), and the Poincaré–Bendixson Theorem (see, e.g., [90]), we conclude that, given a point $(\tilde{u}, \tilde{v}) \in \mathcal{M}$ in the interior of $\mathcal{A}_4$, the α-limit set of $(\tilde{u}, \tilde{v})$, which we denote by $\alpha(\tilde{u}, \tilde{v})$, exists, and

5.1 Characterization of $\mathcal{M}$ When $ac \neq 1$

$$\alpha(\tilde{u}, \tilde{v}) \text{ can be either an equilibrium} \\ \text{or a union of (finitely many) equilibria} \tag{5.20}$$
and non-closed orbits connecting these equilibria.

We stress that, being $(\tilde{u}, \tilde{v})$ in the interior of $\mathcal{A}_4$, we have that

$$\tilde{u} < u_s \ \vee \ \tilde{v} < v_s. \tag{5.21}$$

Now, we observe that

$$\alpha(\tilde{u}, \tilde{v}) \text{ cannot contain the saddle point } (u_s, v_s). \tag{5.22}$$

Indeed, suppose by contradiction that $\alpha(\tilde{u}, \tilde{v})$ does contain (u_s, v_s). Then, we denote by

$$\phi_{(\tilde{u}, \tilde{v})}(t) = \big(u_{(\tilde{u}, \tilde{v})}(t), v_{(\tilde{u}, \tilde{v})}(t)\big)$$

the solution of (2.1) with $\phi_{(\tilde{u}, \tilde{v})}(0) = (\tilde{u}, \tilde{v})$, and we have that there exists a sequence $t_j \to -\infty$ such that $\phi_{(\tilde{u}, \tilde{v})}(t_j)$ converges to (u_s, v_s) as $j \to +\infty$. In particular, in light of (5.21), there exists j_0 sufficiently large such that

$$u_{(\tilde{u}, \tilde{v})}(0) = \tilde{u} < u_{(\tilde{u}, \tilde{v})}(t_{j_0}) \quad \text{or} \quad v_{(\tilde{u}, \tilde{v})}(0) = \tilde{v} < v_{(\tilde{u}, \tilde{v})}(t_{j_0}).$$

Consequently, there exists $t_\star \in (t_{j_0}, 0)$ such that

$$\dot{u}_{(\tilde{u}, \tilde{v})}(t_\star) < 0 \quad \text{or} \quad \dot{v}_{(\tilde{u}, \tilde{v})}(t_\star) < 0.$$

As a result, it follows that $\phi_{(\tilde{u}, \tilde{v})}(t_\star) \notin \mathcal{A}_4$. This, together with the fact that $\phi_{(\tilde{u}, \tilde{v})}(0) \in \mathcal{A}_4$, is in contradiction with (5.19), and the proof of (5.22) is thereby complete.

Thus, from (5.20) and (5.22), we deduce that $\alpha_{(\tilde{u}, \tilde{v})} = \{(0, 0)\}$. This gives that $(0, 0)$ lies on the stable manifold $\mathcal{M}$, and therefore the proof of (5.18) is complete.

Now, we show that

$$\text{there exists } (u_\mathcal{M}, v_\mathcal{M}) \in \partial\big([0, 1] \times [0, 1]\big) \\ \text{such that } \gamma(u_\mathcal{M}) = v_\mathcal{M}. \tag{5.23}$$

To prove it, we first observe that

$$\text{nonfixed point orbits converging to } (u_s, v_s) \\ \text{are not contained in } \mathcal{A}_2. \tag{5.24}$$

Indeed, we suppose by contradiction that there exists a nontrivial orbit contained in $\mathcal{A}_2$ converging to (u_s, v_s). We remark that, in this case, the orbit contained in $\mathcal{A}_2$ cannot be a close orbit, because $\dot{u}$ and $\dot{v}$ have a sign in $\mathcal{A}_2$.

Then, by the Poincaré–Bendixson Theorem (see, e.g., [90]), we conclude that, given a point $(\tilde{u}, \tilde{v}) \in \mathcal{M}$ in the interior of $\mathcal{A}_2$, the α-limit set of $(\tilde{u}, \tilde{v})$, which we denote by $\alpha(\tilde{u}, \tilde{v})$, exists, and it is an equilibrium or a union of (finitely many) equilibria and non-closed orbits connecting these equilibria.

We notice that the set $\alpha(\tilde{u}, \tilde{v})$ cannot contain $(0, 1)$ or $(0, 0)$, since they lay outside $\mathcal{A}_2$.

So, since (u_s, v_s) is the only equilibria, the α-limit set must coincide with it, and the orbit must be a homoclinic. This is in contradiction with (5.11), proving (5.24).

Now, we observe that the inward pointing normal derivative at every point in $\mathcal{A}_2 \cap \mathcal{A}_3 \setminus \{(u_s, v_s)\}$ is negative, since $\dot{u} = 0$ and $\dot{v} < 0$. Hence, no trajectory can enter from this side (see Remark 4.5).

Also, the inward pointing normal derivative at every point in $\mathcal{A}_1 \cap \mathcal{A}_2 \setminus \{(u_s, v_s)\}$ is negative, since $\dot{u} > 0$ and $\dot{v} = 0$. Hence, no trajectory can enter from this side either.

These observations and (5.24) give the desired result in (5.23), and thus Proposition 5.1 is established in the case $ac < 1$.

Now we treat the case $ac > 1$, using the same ideas. In this setting, $\mathcal{M}$ is the stable manifold associated with the saddle point $(0, 0)$. We point out that, in this case, for all points in $[0, 1] \times [0, 1]$, we have that $\dot{u} \leqslant 0$.

Hence, the curve of points satisfying $\dot{v} = 0$, which was also given in (5.2), divides the square $[0, 1] \times [0, 1]$ into two regions $\mathcal{A}_1$ and $\mathcal{A}_2$, defined in (5.9).

Now, one can repeat verbatim the arguments in *Step 1* with obvious modifications, to find that $\mathcal{M} \subset \mathcal{A}_2$.

Since the derivatives of u and v have a sign in $\mathcal{A}_2$ and the set $\mathcal{M}$ in this case is the trajectory of a point converging to $(0, 0)$, the set $\mathcal{M}$ can be represented globally as the graph of a smooth increasing function $\gamma : U \to [0, 1]$ for a suitable interval $U \subseteq [0, 1]$ containing the origin.

As a consequence, the condition $\gamma(0) = 0$ is trivially satisfied in this setting. The existence of a suitable $(u_\mathcal{M}, v_\mathcal{M})$ can be derived reasoning as in *Step 3* with obvious modifications.

Now, we prove that

$$\text{at } u = 0 \text{ the function } \gamma \text{ is tangent} \\ \text{to the line } (\rho - 1 + ac)v - au = 0. \tag{5.25}$$

For this, we recall (5.3), and we see, by inspection, that the Jacobian matrix $J(0, 0)$ has two eigenvectors, namely $(0, 1)$ and $(\rho - 1 + ac, a)$. The first one is tangent to the line $u = 0$, which is the unstable manifold of $(0, 0)$, as one can easily verify.

Thus, the second eigenvector is the one tangent to $\mathcal{M}$, as prescribed by the Stable Manifold Theorem (see, e.g., [77]).

Hence, in $(0, 0)$ the manifold $\mathcal{M}$ is tangent to the line

$$(\rho - 1 + ac)v - au = 0$$

and so is the function γ in $u = 0$. This proves (5.25), and thus Proposition 5.1 is established in the case $ac > 1$ as well. □

5.2 Characterization of $\mathcal{M}$ When $ac = 1$

Here we will prove the counterpart of Proposition 5.1 in the degenerate case $ac = 1$.

To this end, looking at the velocity fields, we first observe that

$$\text{trajectories starting in } (0, 1) \times (-\infty, 1) \text{ at time } t = 0 \tag{5.26}$$
$$\text{remain in } (0, 1) \times (-\infty, 1) \text{ for all time } t > 0.$$

We also point out that

$$\text{trajectories entering the region}$$
$$\mathcal{R} := \{u \in (0, 1), u + v < 0\} \quad \text{at some time } t_0 \in \mathbb{R}, \tag{5.27}$$
$$\text{remain in that region for all time } t > t_0,$$

since

$$\dot{v} = \rho v(1 - u - v) - au = -\rho u - au < 0$$

along $\{u \in (0, 1), u + v = 0\}$ (see Remark 4.5).

Also, by the Center Manifold Theorem (see, e.g., Theorem 1 on page 16 of [18] or pages 89–90 in [83]), there exists a collection $\mathcal{M}_0$ of invariant curves, which are all tangent at the origin to the eigenvector corresponding to the null eigenvalue, that is, the straight line $\rho v - au = 0$.

Then, we define

$$\mathcal{M} := \mathcal{M}_0 \cap ([0, 1] \times [0, 1]),$$

and we observe that this intersection is nonvoid, given the tangency property of $\mathcal{M}_0$ at the origin.

In what follows, for every $t \in \mathbb{R}$, we denote by

$$(u(t), v(t)) = \phi_p(t)$$

the orbit of $p \in \mathcal{M} \setminus \{(0,0)\}$. We start by providing an observation related to negative times:

Lemma 5.4 *Suppose that $ac = 1$. If $p \in \mathcal{M} \setminus \{(0,0)\}$, then $\phi_p(t)$ cannot approach the origin for negative values of t.*

Proof We argue by contradiction and denote by $t_1, \ldots, t_n, \ldots$ a sequence of such negative values of t, for which $t_n \to -\infty$ and

$$\lim_{n \to +\infty} \phi_p(t_n) = (0, 0).$$

Up to a subsequence, we can also suppose that

$$u(t_{n+1}) < u(t_n). \tag{5.28}$$

In light of (5.27), we have that, for all $T \leqslant 0$,

$$\phi_p(T) \notin \mathcal{R}. \tag{5.29}$$

Indeed, if $\phi_p(T) \in \mathcal{R}$, we deduce from (5.27) that $\phi_p(t) \in \mathcal{R}$ for all $t \geqslant T$. In particular, we can take $t = 0 \geqslant T$ and conclude that $p = \phi_p(0) \in \mathcal{R}$, and this is in contradiction with the assumption that $p \in \mathcal{M} \setminus \{(0,0)\}$.

As a by-product of (5.29), we obtain that, for all $T \leqslant 0$,

$$\phi_p(T) \in \{u \in (0,1),\ u+v \geqslant 0\} \subseteq \{\dot{u} = -u(u+v) \leqslant 0\}.$$

In particular

$$u(t_n) - u(t_{n+1}) = \int_{t_{n+1}}^{t_n} \dot{u}(\tau)\, d\tau \leqslant 0,$$

which contradicts (5.28), and consequently we have established the desired result. □

Now we show that the ω-limit set of any point lying on the global center manifold coincides with the origin, according to the next result:

Lemma 5.5 *Suppose that $ac = 1$. If $p \in \mathcal{M}$, then its ω-limit is $\{(0,0)\}$.*

Proof We observe that, for every $t > 0$,

$$\phi_p(t) \in [0,1] \times [0,1]. \tag{5.30}$$

5.2 Characterization of $\mathcal{M}$ When $ac = 1$

By Remark 3.1, the other possibility would be $T_s(p) < +\infty$. Therefore, to prove (5.30), we suppose, by contradiction, that there exists $t_0 \geq 0$ such that $\phi_p(t_0) \in [0, 1] \times \{0\}$, that is, $v(t_0) = 0$.

Since $(0, 0)$ is an equilibrium, it follows that $u(t_0) \neq 0$. In particular, $u(t_0) > 0$ and accordingly

$$\dot{v}(t_0) = -au(t_0) < 0.$$

This means that $v(t_0 + \varepsilon) < 0$ for all $\varepsilon \in (0, \varepsilon_0)$ for a suitable $\varepsilon_0 > 0$. Looking again at the velocity fields, this entails that $\phi_p(t) \in (0, 1) \times (-\infty, 0)$ for all $t > \varepsilon_0$.

Consequently, $\phi_p(t)$ cannot approach the straight line $\rho v - au = 0$ for $t > \varepsilon_0$.

This, combined with Lemma 5.4, says that the trajectory emanating from p can never approach the straight line $\rho v - au = 0$ at the origin, in contradiction with the definition of $\mathcal{M}$, and thus the proof of (5.30) is complete.

From (5.30) and the Poincaré–Bendixson Theorem (see, e.g., [90]), we deduce that the ω-limit of p can be either a cycle or an equilibrium or a union of (finitely many) equilibria and non-closed orbits connecting these equilibria. We observe that the ω-limit of p cannot be a cycle, since $\dot{u}$ has a sign in $[0, 1] \times [0, 1]$.

Moreover, it cannot contain the sink $(0, 1)$, due to Lemma 5.4. Hence, the only possibility is that the ω-limit of p coincides with $(0, 0)$, which is the desired result.

We also remark that the α-limit of p cannot be a cycle in $[0, 1] \times [0, 1]$ since $\dot{u}$ has a sign. Moreover, it cannot contain $(0, 1)$, which is a sink. If the orbit of p is all contained in $[0, 1] \times [0, 1]$, then the α-limit set cannot contain $(0, 0)$, since this would generate a close trajectory. Therefore, the orbit of $\phi_p(t)$ must intersect the complementary set of $[0, 1] \times [0, 1]$. □

As a consequence of Lemma 5.5 and the fact that $\dot{u} < 0$ in $(0, 1] \times [0, 1]$, we obtain the following statement:

Corollary 5.6 *For $ac = 1$, every trajectory in $\mathcal{M}$ has the form $\{\phi_p(t), t \in \mathbb{R}\}$, with*

$$\lim_{t \to +\infty} \phi_p(t) = (0, 0),$$

and there exists $t_p \in \mathbb{R}$ such that

$$\phi_p(t_p) \in \big(\{1\} \times [0, 1]\big) \cup \big([0, 1] \times \{1\}\big).$$

The result in Corollary 5.6 can be sharpened in view of the following statement (which can be seen as the counterpart of Proposition 5.1 in the degenerate case $ac = 1$): Namely, since the center manifold can in principle contain many different trajectories (see, e.g.,

Figure 5.3 in [18]), we provide a tailor-made argument that excludes this possibility in the specific case that we deal with.

Proposition 5.7 *For $ac = 1$, the set $\mathcal{M}$ contains one, and only one, orbit, which is asymptotic to the origin as $t \to +\infty$, and that can be written as a graph*

$$\gamma : [0, u_{\mathcal{M}}] \to [0, v_{\mathcal{M}}],$$

for some

$$(u_{\mathcal{M}}, v_{\mathcal{M}}) \in \big(\{1\} \times [0, 1]\big) \cup \big((0, 1] \times \{1\}\big),$$

where γ is an increasing C^2 function such that $\gamma(0) = 0$, $\gamma(u_{\mathcal{M}}) = v_{\mathcal{M}}$ and the graph of γ at the origin is tangent to the line $\rho v - au = 0$.

Proof First of all, we show that

$$\mathcal{M} \text{ contains one, and only one, orbit.} \tag{5.31}$$

Suppose, by contradiction, that $\mathcal{M}$ contains two different orbits, which we denote by $\mathcal{M}_-$ and $\mathcal{M}_+$. Using Corollary 5.6, we can suppose that $\mathcal{M}_+$ lies above $\mathcal{M}_-$ and

$$\begin{aligned}&\text{the region } \mathcal{P} \subset [0, 1] \times [0, 1] \text{ contained} \\ &\text{between } \mathcal{M}_+ \text{ and } \mathcal{M}_- \text{ lies in } \{\dot{u} < 0\}.\end{aligned} \tag{5.32}$$

Consequently, for every $p \in \mathcal{P}$, it follows that

$$\lim_{t \to +\infty} \phi_p(t) = (0, 0). \tag{5.33}$$

In particular, we can take an open ball $B \subset \mathcal{P}$ in the vicinity of the origin, denote by $\mu(t)$ the Lebesgue measure of

$$\mathcal{S}(t) := \{\phi_p(t), \ p \in B\},$$

and write that $\mu(0) > 0$ and

$$\lim_{t \to +\infty} \mu(t) = 0. \tag{5.34}$$

We point out that $\mathcal{S}(t)$ lies in the vicinity of the origin for all $t \geq 0$, thanks to (5.32). As a consequence, for all $t, \tau > 0$, changing variable

5.2 Characterization of $\mathcal{M}$ When $ac = 1$

$$y := \phi_x(\tau) = x + \int_0^\tau \frac{d\phi_x(\theta)}{d\theta} d\theta = x + \tau \frac{d\phi_x(0)}{dt} + O(\tau^2),$$

we find that

$$\mu(t+\tau) = \int_{\mathcal{S}(t+\tau)} dy$$

$$= \int_{\mathcal{S}(t)} \left| \det \left(D_x \phi_x(\tau) \right) \right| dx$$

$$= \int_{\mathcal{S}(t)} \left| \det D_x \left(x + \tau \frac{d\phi_x(0)}{dt} + O(\tau^2) \right) \right| dx$$

$$= \int_{\mathcal{S}(t)} \left(1 + \tau \operatorname{Tr} \left(D_x \frac{d\phi_x(0)}{dt} \right) + O(\tau^2) \right) dx$$

$$= \mu(t) + \tau \int_{\mathcal{S}(t)} \operatorname{Tr} \left(D_x \frac{d\phi_x(0)}{dt} \right) dx + O(\tau^2),$$

where Tr denotes the trace of a (2×2)-matrix.

As a consequence,

$$\frac{d\mu}{dt}(t) = \int_{\mathcal{S}(t)} \operatorname{Tr} \left(D_x \frac{d\phi_x(0)}{dt} \right) dx. \tag{5.35}$$

Also, using the notation $x = (u, v)$, we can write (2.1) when $ac = 1$ in the form

$$\frac{d\phi_x}{dt}(t) = \dot{x}(t) = \begin{pmatrix} \dot{u}(t) \\ \dot{v}(t) \end{pmatrix} = \begin{pmatrix} -u(t)(u(t) + v(t)) \\ \rho v(t)(1 - u(t) - v(t)) - au(t) \end{pmatrix}.$$

Accordingly,

$$D_x \frac{d\phi_x(0)}{dt}$$

$$= \begin{pmatrix} -\partial_u \bigl(u(u+v)\bigr) & -\partial_v \bigl(u(u+v)\bigr) \\ \partial_u \bigl(\rho v(1-u-v) - au\bigr) & \partial_v \bigl(\rho v(1-u-v) - au\bigr) \end{pmatrix},$$

whence

$$\operatorname{Tr}\left(D_x \frac{d\phi_x(0)}{dt}\right)$$
$$= -\partial_u\big(u(u+v)\big) + \partial_v\big(\rho v(1-u-v) - au\big) \qquad (5.36)$$
$$= -2u - v + \rho(1-u-v) - \rho v$$
$$= \rho + O(|x|)$$

for x near the origin.

As a result, recalling (5.33), we can take t sufficiently large, such that $\mathcal{S}(t)$ lies in a neighborhood of the origin, and exploit (5.36) to write that

$$\operatorname{Tr}\left(D_x \frac{d\phi_x(0)}{dt}\right) \geqslant \frac{\rho}{2}$$

and then (5.35) to conclude that

$$\frac{d\mu}{dt}(t) \geqslant \frac{\rho}{2} \int_{\mathcal{S}(t)} dx = \frac{\rho}{2} \mu(t).$$

This implies that $\mu(t)$ diverges (exponentially fast) as $t \to +\infty$, which is in contradiction with (5.34). The proof of (5.31) is thereby complete.

Now, we check the other claims in the statement of Proposition 5.7. The asymptotic property as $t \to +\infty$ is a consequence of Corollary 5.6. Also, the graphical property and the monotonicity property of the graph follow from the fact that $\mathcal{M} \subset \{\dot{u} < 0\}$. The smoothness of the graph follows from the smoothness of the center manifold. The fact that $\gamma(0) = 0$ and $\gamma(u_{\mathcal{M}}) = v_{\mathcal{M}}$ follow also from Corollary 5.6. The tangency property at the origin is a consequence of the tangency property of the center manifold to the center eigenspace. □

As a by-product of the proof of Proposition 5.7, we also obtain the following information:

Corollary 5.8 *Suppose that $ac = 1$. Then, we have that $\dot{u} \leqslant 0$ in $[0,1] \times [0,1]$, and the curve (5.2) divides the square $[0,1] \times [0,1]$ into two regions $\mathcal{A}_1$ and $\mathcal{A}_2$, defined in (5.9).*

Furthermore, the sets $\mathcal{A}_1$ and $\mathcal{A}_2$ are separated by the curve $\dot{v} = 0$, given by the graph of the continuous function σ given in (5.5).

Finally, there holds that

$$\mathcal{M} \subset \mathcal{A}_2. \qquad (5.37)$$

5.3 Study of the Dynamics

We observe that, by the Stable Manifold Theorem and the Center Manifold Theorem, the statement in (v) of Theorem 3.2 is obviously fulfilled.

Hence, to complete the proof of Theorem 3.2, it remains to show that the statement in (iv) holds true. To this aim, exploiting the useful pieces of information in Propositions 5.1 and 5.7, we first give a characterization of the sets $\mathcal{E}$ and $\mathcal{B}$:

Proposition 5.9 *The sets in (3.3) and (3.4) are characterized by*

$$\mathcal{E} = \{(u, v) \in [0, u_\mathcal{M}] \times [0, 1] \text{ s.t. } v < \gamma(u)\} \\ \cup \left((u_\mathcal{M}, 1] \times [0, 1]\right) \tag{5.38}$$

and

$$\mathcal{B} = \{(u, v) \in [0, u_\mathcal{M}] \times [0, 1] \text{ s.t. } v > \gamma(u)\}, \tag{5.39}$$

where γ is the parametrization of $\mathcal{M}$, as given by Propositions 5.1 (when $ac \neq 1$) and 5.7 (when $ac = 1$).

In (5.38) we use the convention that $(u_\mathcal{M}, 1] \times [0, 1] = \emptyset$ in the case $u_\mathcal{M} = 1$. The sets $\mathcal{E}$ and $\mathcal{B}$ can be visualized in two particular cases in Fig. 5.2.

Proof of Proposition 5.9 We let γ be the parametrization of $\mathcal{M}$, as given by Propositions 5.1 (when $ac \neq 1$) and 5.7 (when $ac = 1$). Let us call $\mathcal{X}$ and $\mathcal{Y}$ the sets in (5.38) and (5.39), that is,

$$\mathcal{X} := \{(u, v) \in [0, u_\mathcal{M}] \times [0, 1] \text{ s.t. } v < \gamma(u)\} \\ \cup \left((u_\mathcal{M}, 1] \times [0, 1]\right),$$
$$\text{and} \quad \mathcal{Y} := \{(u, v) \in [0, u_\mathcal{M}] \times [0, 1] \text{ s.t. } v > \gamma(u)\}$$

(the second set in the definition of $\mathcal{X}$ is understood to be $\emptyset$ if $u_\mathcal{M} = 1$). The goal is to prove that $\mathcal{X} \equiv \mathcal{E}$ and $\mathcal{Y} \equiv \mathcal{B}$. We recall from Propositions 5.1 and 5.7 that $(\mathcal{X}, \mathcal{Y}, \mathcal{M})$ is a partition of $[0, 1] \times [0, 1]$. Hence, since the sets $\mathcal{E}$, $\mathcal{B}$, and $\mathcal{M}$ are disjoint, if we show that $\mathcal{X} \subseteq \mathcal{E}$ and $\mathcal{Y} \subseteq \mathcal{B}$, we are done.

We first deal with the inclusion $\mathcal{X} \subseteq \mathcal{E}$. Namely, recalling (3.4), we will show that

$$\text{all } (u_0, v_0) \in \mathcal{X} \text{ have } T_s(u_0, v_0) < +\infty. \tag{5.40}$$

For this, we first notice that, gathering together (5.6), (5.7), (5.8), and (5.11), we find that

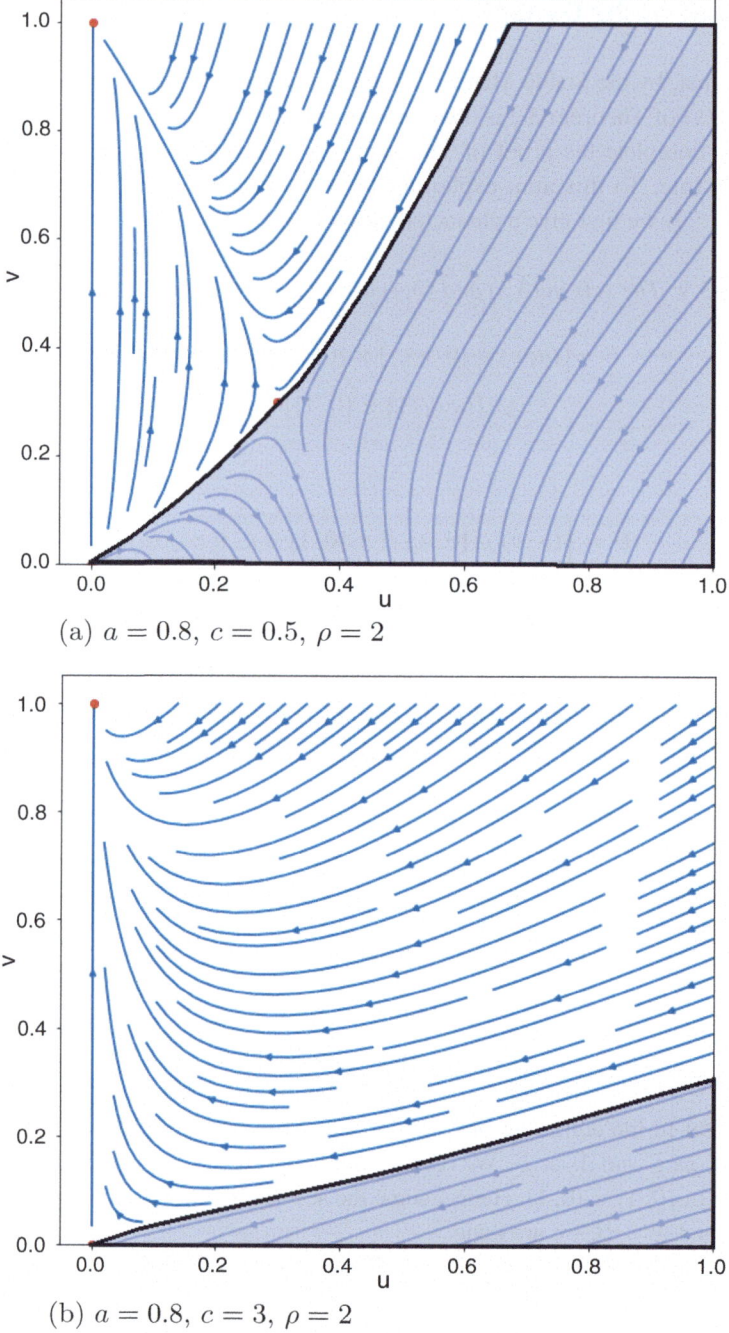

Fig. 5.2 The figures show the phase portrait for the indicated values of the coefficients. Plotted in blue are the orbits of the points. The red dots show the equilibria. Plotted in violet is the set $\mathcal{E}$

5.3 Study of the Dynamics

$$\text{no closed orbit exists in } [0, 1] \times [0, 1] \tag{5.41}$$

(in the case $0 < ac < 1$, and this holds true in the case $ac \geqslant 1$ where $\dot{u}$ has a sign).
In addition,

$$\text{the } \omega\text{-limit of any point in } \mathcal{X}, \text{ if it exists,} \tag{5.42}$$
$$\text{does not contain equilibria.}$$

Indeed, by Propositions 5.1 (when $ac \neq 1$) and 5.7 (when $ac = 1$), we have that $\gamma(0) = 0 < 1$, and therefore $(0, 1) \notin \overline{\mathcal{X}}$. Moreover, if $ac < 1$, a trajectory in $\mathcal{X}$ cannot converge to (u_s, v_s), since $\mathcal{X}$ does not contain points of the corresponding stable manifold $\mathcal{M}$, nor to $(0, 0)$, since this is a repulsive equilibrium, cf. Theorem 3.2. If instead $ac \geqslant 1$, then trajectories cannot converge to $(0, 0)$, since $\mathcal{X}$ does not contain points of $\mathcal{M}$, which in this case coincides with the stable or center manifold of $(0, 0)$. These observations complete the proof of (5.42).

Moreover, by Lemma 4.7, no trajectory can exit $\mathcal{X}$, since $\partial \mathcal{X} \setminus \partial([0, 1] \times [0, 1])$ coincide with a trajectory (which contains also the point at the boundary $(u_\mathcal{M}, \gamma(\mathcal{M}))$).

From the latter observation, (5.41), (5.42), and the Poincaré–Bendixson Theorem (see, e.g., [90]), we have that every trajectory with initial point

$$(u_0, v_0) \in \mathcal{X}$$

cannot remain in $[0, 1] \times [0, 1]$ for all $t > 0$, that is, $T_s(u_0, v_0) < +\infty$. This shows that $\mathcal{X} \subseteq \mathcal{E}$.

We now claim that

$$\big((u_\mathcal{M}, 1] \times [0, 1]\big) \subseteq \mathcal{E}. \tag{5.43}$$

To this end, we observe that there are neither cycles nor equilibria in $(u_\mathcal{M}, 1] \times [0, 1]$, and therefore we can use the Poincaré–Bendixson Theorem (see, e.g., [90]) to conclude that for any point $(u_0, v_0) \in (u_\mathcal{M}, 1] \times [0, 1]$, its ω-limit set, if it exists, must satisfy

$$\omega(u_0, v_0) \cap ((u_\mathcal{M}, 1] \times [0, 1]) = \varnothing. \tag{5.44}$$

Now, a trajectory can exit $(u_\mathcal{M}, 1] \times [0, 1]$ only from the side $\{u_\mathcal{M}\} \times (0, 1)$ and entering the set $\mathcal{X}$, and therefore (5.43) is a consequence of (5.40) in this case.

Thanks to (5.44), the only other possibility is that $T_s(u_0, v_0) < +\infty$, giving directly (5.43).

From (5.40) and (5.43), we obtain (5.38), as desired.

We now prove (5.39); namely we show that

for all $(u_0, v_0) \in \mathcal{Y}$ we have that

$$(u(t), v(t)) \to (0, 1) \text{ as } t \to +\infty. \tag{5.45}$$

Hence, again by Lemma 4.7, we have that no trajectory exits $\mathcal{Y}$, since no trajectory can cross $\mathcal{M}$.

Also,

$$\mathcal{Y} \cap ((0, 1) \times \{0\}) = \varnothing.$$

Thus, for $(u_0, v_0) \in \mathcal{Y}$, it must be $T_s(u_0, v_0) = +\infty$ and

$$\omega(u_0, v_0) \subset \mathcal{Y}. \tag{5.46}$$

Let us now investigate $\omega(u_0, v_0)$. To this end, we observe that (u_s, v_s) (if $0 < ac < 1$) and $(0, 0)$ are not in $\mathcal{Y}$. Moreover, no trajectory starting in $\mathcal{Y}$ converges to (u_s, v_s) (if $0 < ac < 1$), nor to $(0, 0)$, since $\mathcal{Y}$ does not contain points on $\mathcal{M}$.

In addition, recalling (5.41), we have that there are no limit cycles in $\mathcal{Y}$. As a consequence, by the Poincaré–Bendixson Theorem (see, e.g., [90]), we have that every trajectory starting in $\mathcal{Y}$ converges to $(0, 1)$, proving (5.45). Hence, the proof of (5.39) is complete as well. $\square$

With this, we are now able to complete the proof of Theorem 3.2:

Proof of (iv) of Theorem 3.2 The statement in (iv) of Theorem 3.2 is a direct consequence of the parametrization of the manifold $\mathcal{M}$, as given by Proposition 5.1 for $ac \neq 1$ and by Proposition 5.7 for $ac = 1$, and the characterization of the sets $\mathcal{B}$ and $\mathcal{E}$, as given by Proposition 5.9. $\square$

6 Parameters Dependence

In this chapter we discuss the dependence on the parameters involved in the system (2.1).

The dynamics of the system in (2.1) depends qualitatively only on ac, but of course the position of the saddle equilibrium and the size and shape of the basins of attraction depend quantitatively upon all the parameters. Here we perform a deep analysis on each parameter separately.

We notice that the system in (2.1) does not present a variational structure, due to the presence of the terms $-acu$ in the first equation and $-au$ in the second one, which are of first order in u. Thus, the classical methods of the calculus of variations cannot be used, and we have to make use of ad hoc arguments, of geometrical flavor.

6.1 Dependence on the Parameter c

We start by studying the dependence on c, which represents the losses (soldier death and missing births) caused by the war for the first population with respect to the second one. In the following proposition, we will express the dependence on c of the basin of attraction $\mathcal{E}$ in (3.4) by writing explicitly $\mathcal{E}(c)$.

Proposition 6.1 (Dependence of the Dynamics On c) *With the notation in (3.4), we have that:*

(i) *If $0 < c_1 < c_2$, then $\mathcal{E}(c_2) \subset \mathcal{E}(c_1)$.*
(ii) *It holds that*

$$\bigcap_{c>0} \mathcal{E}(c) = (0, 1] \times \{0\}. \tag{6.1}$$

We remark that the behavior for c small is included by (i) of Theorem 3.2: In this case, there is a saddle point $(u_s, v_s) \in (0, 1) \times (0, 1)$, and for all $c > 0$ we get $\mathcal{E}(c) \neq (0, 1] \times [0, 1]$. We do not investigate the case $c = 0$, which results in a special case where the saddle (u_s, v_s) and the sink $(0, 1)$ collapse in a point with one negative and one zero eigenvalue.

On the other hand, as $c \to +\infty$, the set $\mathcal{E}(c)$ gets smaller and smaller until the first population has no chances of victory if the second population has a positive size.

As one would expect, Proposition 6.1 tells us that the greater the cost of the war for the first population, the fewer possibilities of victory there are for it.

Proof of Proposition 6.1

(i) We take $c_2 > c_1 > 0$. Now, in the notation of Propositions 5.1 (if $ac \neq 1$) and 5.7 (if $ac = 1$), thanks to the characterization in (5.38), the inclusion in (i) reduces to

$$\gamma_{c_1}(u) > \gamma_{c_2}(u) \quad \text{for any } u \in (0, u_\mathcal{M}^{c_1}]. \tag{6.2}$$

Notice that this gives us automatically $u_\mathcal{M}^{c_1} < u_\mathcal{M}^{c_2}$, and the inclusion

$$(\mathcal{E}(c_2) \cap (u_\mathcal{M}^{c_1}, 1] \times [0, 1]) \subset (\mathcal{E}(c_1) \cap (u_\mathcal{M}^{c_1}, 1] \times [0, 1])$$

is trivial.

Suppose by the absurd that (6.2) is not true; hence there is some value $\bar{u} \in (0, u_\mathcal{M}^{c_1}]$ such that $\gamma_{c_1}(\bar{u}) < \gamma_{c_2}(\bar{u})$. Let us now consider a value $\bar{v} \in (\gamma_{c_1}(\bar{u}), \gamma_{c_2}(\bar{u}))$. Then, $(\bar{u}, \bar{v}) \in \mathcal{E}(c_2) \setminus \mathcal{E}(c_1)$. Observe that $(0, 1] \times \{0\} \subset \mathcal{E}(c_1)$ and $T_s(\bar{u}, \bar{v}) < +\infty$. In particular,

$$\begin{array}{c} \text{there must be a time } T \in (0, T_s(\bar{u}, \bar{v})) \\ \text{such that the trajectory } \phi_{(\bar{u},\bar{v})}^{c_2}(t) \text{ enters } \mathcal{E}(c_1). \end{array} \tag{6.3}$$

Moreover, by Remark 3.1 (see also Remark 4.2), the trajectory $\phi_{(\bar{u},\bar{v})}^{c_2}(t)$ does not leave $[0, 1] \times [0, 1]$ for $t \in (0, T_s(\bar{u}, \bar{v}))$; therefore it enters $\mathcal{E}(c_1)$ from the side given by $\mathcal{M}(c_1)$.

Let us now compute the normal derivative to $\mathcal{M}(c_1)$ at the point $(\tilde{u}, \tilde{v}) = \phi_{(\bar{u},\bar{v})}^{c_2}(T)$. We use the notation

$$\dot{u}_1 := \tilde{u}(1 - ac_1 - \tilde{u} - \tilde{v}),$$

$$\dot{u}_2 := \tilde{u}(1 - ac_2 - \tilde{u} - \tilde{v}),$$

and $\quad \dot{v} := \rho\tilde{v}(1 - \tilde{u} - \tilde{v}) - a\tilde{u}.$

Since $\mathcal{M}(c_1)$ is the graph of γ_{c_1} and by the properties of $\mathcal{M}$ given in Corollary 5.2 for $0 < ac < 1$, Corollary 5.3 for $ac > 1$, and Proposition 5.7 for $ac = 1$, the normal vector ν to $\mathcal{M}(c_1)$ at $(\tilde{u}, \tilde{v})$ is given by:

6.1 Dependence on the Parameter c

1. If $\dot{v} > 0$,
$$\nu := \left(\frac{-\dot{v}}{\|(\dot{u}_1, \dot{v})\|}, \frac{\dot{u}_1}{\|(\dot{u}_1, \dot{v})\|} \right).$$

2. If $\dot{v} < 0$,
$$\nu := \left(\frac{\dot{v}}{\|(\dot{u}_1, \dot{v})\|}, \frac{-\dot{u}_1}{\|(\dot{u}_1, \dot{v})\|} \right).$$

3. If $\dot{v} = 0$ (i.e., when we are at $(u_s^{c_1}, v_s^{c_1})$ for $0 < ac < 1$), by Proposition 5.1,
$$\nu := \frac{1}{\sqrt{1+c_1^2}} (-1, c_1).$$

Now, for each case, we compute the product between the normal vector and the direction $(\dot{u}_2, \dot{v})$ of the trajectory $\phi_{(\tilde{u}, \tilde{v})}^{c_2}(t)$ at $(\tilde{u}, \tilde{v})$. We get:

1. If $\dot{v} > 0$, then
$$\frac{1}{\|(\dot{u}_1, \dot{v})\|} \dot{v}(\dot{u}_1 - \dot{u}_2) = \frac{a\tilde{u}\tilde{v}(c_2 - c_1)}{\|(\dot{u}_1, \dot{v})\|} > 0.$$

2. If $\dot{v} < 0$, then
$$\frac{1}{\|(\dot{u}_1, \dot{v})\|} \dot{v}(\dot{u}_2 - \dot{u}_1) = \frac{a\tilde{u}\tilde{v}(c_1 - c_2)}{\|(\dot{u}_1, \dot{v})\|} > 0.$$

3. If $\dot{v} = 0$, owning the formula (3.5) for $(u_s^{c_1}, v_s^{c_2})$, then
$$\frac{1}{\sqrt{1+c_1^2}} (-\dot{u}_2 + c\dot{v}) = \frac{1}{\sqrt{1+c_1^2}} \frac{1 - ac_1}{1 + \rho c_1} \rho c_1 (ac_2 - ac_1) > 0.$$

Since the scalar product of the normal to $\mathcal{M}(c_1)$ and the trajectory is always positive, the trajectory cannot enter in $\mathcal{E}(c_1)$ (see Remark 4.5), contradicting (6.3). Hence, (6.2) holds true.

(ii) We first show that for all $\varepsilon > 0$ there exists $c_\varepsilon > 0$ such that for all $c \geqslant c_\varepsilon$ it holds that

$$\mathcal{E}(c) \subset \{(u, v) \in [0, 1] \times [0, 1] \text{ s.t. } v < \varepsilon u\}. \tag{6.4}$$

The inclusion in (6.4) is also equivalent to

$$\{(u, v) \in [0, 1] \times [0, 1] \text{ s.t. } v > \varepsilon u\} \subset \mathcal{B}(c), \tag{6.5}$$

and the strict inequality is justified by the fact that $\mathcal{E}(c)$ and $\mathcal{B}(c)$ are separated by $\mathcal{M}$, according to Proposition 5.9. We now establish the inclusion in (6.5). For this, let

$$\mathcal{T}_\varepsilon := \{(u, v) \in [0, 1] \times [0, 1] \text{ s.t. } v \geqslant \varepsilon u\}. \tag{6.6}$$

Now, we can choose c large enough such that the condition $ac \geqslant 1$ is fulfilled. In this way, thanks to (ii) and (iii) of Theorem 3.2, the only equilibria are the points $(0, 0)$ and $(0, 1)$.

Now, the component of the velocity in the inward normal direction to $\mathcal{T}_\varepsilon$ on the side $\{v = \varepsilon u\}$ is given by

$$(\dot u, \dot v) \cdot \frac{(-\varepsilon, 1)}{\sqrt{1+\varepsilon^2}} = \frac{\dot v - \varepsilon \dot u}{\sqrt{1+\varepsilon^2}}$$

$$= \frac{1}{\sqrt{1+\varepsilon^2}}\bigl(\rho v(1-u-v) - au - \varepsilon u(1-u-v) + \varepsilon acu\bigr)$$

$$= \frac{1}{\sqrt{1+\varepsilon^2}}\bigl[(\rho v - \varepsilon u)(1-u-v) + (\varepsilon c - 1)au\bigr]$$

$$= \frac{1}{\sqrt{1+\varepsilon^2}}\bigl[(\rho \varepsilon u - \varepsilon u)(1-u-\varepsilon u) + (\varepsilon c - 1)au\bigr],$$

which is positive for

$$c > c_\varepsilon := \frac{2\varepsilon(1+\rho) + a}{\varepsilon a}. \tag{6.7}$$

This says that no trajectory in $\mathcal{T}_\varepsilon$ can exit $\mathcal{T}_\varepsilon$ from the side $\{v = \varepsilon u\}$ (see Remark 4.2).

The other parts of $\partial \mathcal{T}_\varepsilon$ belong to $\partial((0,1) \times (0,1))$ but not to $[0,1] \times \{0\}$. As a consequence, no trajectory can exit $\mathcal{T}_\varepsilon$, so

every trajectory in $\mathcal{T}_\varepsilon$ is well defined for all $t \geqslant 0$ and belongs to $\mathcal{T}_\varepsilon$. (6.8)

From this, (5.41), and the Poincaré–Bendixson Theorem (see [90]), we conclude that the ω-limit of any trajectory starting in $\mathcal{T}_\varepsilon$ can be either an equilibrium or a union of (finitely many) equilibria and non-closed orbits connecting these equilibria.

Now, we claim that, possibly taking c larger in (6.7),

$$\mathcal{M} \subset ([0, 1] \times [0, 1]) \setminus \mathcal{T}_\varepsilon. \tag{6.9}$$

Indeed, suppose by contradiction that there exists $(\tilde u, \tilde v) \in \mathcal{M} \cap \mathcal{T}_\varepsilon$. Then, in light of (6.8), a trajectory passing through $(\tilde u, \tilde v)$ and converging to $(0, 0)$ has to be entirely contained in $\mathcal{T}_\varepsilon$.

6.2 Dependence on the Parameter ρ

On the other hand, by Propositions 5.1 and 5.7, we know that at $u = 0$ the manifold $\mathcal{M}$ is tangent to the line $(\rho - 1 + ac)v - au = 0$. Hence, if we choose c large enough such that

$$\frac{a}{\rho - 1 + ac} < \varepsilon,$$

we obtain that this line is below the line $v = \varepsilon u$, thus reaching a contradiction. This establishes (6.9).

From (6.9), we deduce that, given $(\tilde{u}, \tilde{v}) \in \mathcal{T}_\varepsilon$, and denoting $\omega_{(\tilde{u},\tilde{v})}$ the ω-limit of $(\tilde{u}, \tilde{v})$,

$$\omega_{(\tilde{u},\tilde{v})} \neq \{(0, 0)\}, \tag{6.10}$$

provided that c is taken large enough.

Furthermore, $\omega_{(\tilde{u},\tilde{v})}$ cannot consist of the two equilibria $(0, 0)$ and $(0, 1)$ and non-closed orbits connecting these equilibria, due to the fact that $(0, 1)$ is a sink.

As a consequence of this and (6.10), we obtain that $\omega_{(\tilde{u},\tilde{v})} = \{(0, 1)\}$ for any $(\tilde{u}, \tilde{v}) \in \mathcal{T}_\varepsilon$, provided that c is large enough.

Thus, recalling (3.3) and (6.6), this proves (6.5), and therefore (6.4)

Now, using (6.4), we see that for every $\varepsilon > 0$,

$$\bigcap_{c>0} \mathcal{E}(c) \subseteq \mathcal{E}(c_\varepsilon)$$

$$\subseteq \{(u, v) \in [0, 1] \times [0, 1] \text{ s.t. } v < \varepsilon u\}.$$

Accordingly,

$$\bigcap_{c>0} \mathcal{E}(c) \subseteq \bigcap_{\varepsilon>0} \{(u, v) \in [0, 1] \times [0, 1] \text{ s.t. } v < \varepsilon u\}$$

$$= (0, 1] \times \{0\},$$

which gives (6.1), as desired. □

6.2 Dependence on the Parameter ρ

Now we analyze the dependence of the dynamics on the parameter ρ, that is, the fitness of the second population v with respect to the fitness of the first one u.

In the following proposition, we will make it explicit the dependence on ρ by writing $\mathcal{E}(\rho)$ and $\mathcal{B}(\rho)$.

Proposition 6.2 (Dependence of the Dynamics On ρ**)** *With the notation in* (3.3) *and* (3.4), *we have that:*

(i) *When* $\rho = 0$, *for any* $v \in [0, 1]$ *the point* $(0, v)$ *is an equilibrium. If* $v \in (1 - ac, 1]$, *then it corresponds to a strictly negative eigenvalue and a null one. If instead* $v \in [0, 1 - ac)$, *then it corresponds to a strictly positive eigenvalue and a null one. Moreover,*

$$\mathcal{B}(0) = \emptyset, \tag{6.11}$$

and for any $\varepsilon < ac/2$ *and any* $\delta < \varepsilon c/2$ *we have that*

$$[0, 1] \times [0, 1 - ac) \subseteq \mathcal{E}(0) \subseteq \mathcal{T}_{\varepsilon,\delta}, \tag{6.12}$$

where

$$\mathcal{T}_{\varepsilon,\delta} := \{(u, v) \in [0, 1] \times [0, 1] \text{ s.t. } \delta v - \varepsilon u \leqslant \delta(1 - \varepsilon)\}. \tag{6.13}$$

(ii) *For any* $\varepsilon < ac/3$ *and any* $\delta < \varepsilon c/2$, *it holds that*

$$\bigcup_{0 < \rho < a/3} \mathcal{E}(\rho) \subseteq \mathcal{T}_{\varepsilon,\delta},$$

where $\mathcal{T}_{\varepsilon,\delta}$ *is defined in* (6.13).

(iii) *It holds that*

$$\bigcap_{\omega > 0} \bigcup_{\rho > \omega} \mathcal{E}(\rho) = (0, 1] \times \{0\}. \tag{6.14}$$

We point out that the case $\rho = 0$ is not comprehended in Theorem 3.2. As a matter of fact, the dynamics of this case is qualitatively very different from all the other cases. Indeed, for $\rho = 0$ the domain $[0, 1] \times [0, 1]$ is not divided into $\mathcal{E}$ and $\mathcal{B}$, since more attractive equilibria appear on the line $\{0\} \times (0, 1)$. Thus, even if the second population cannot grow, it still has some chance of victory.

As soon as ρ is positive, on the line $u = 0$ only the equilibrium $(0, 1)$ survives, and it attracts all the points that were going to the line $\{0\} \times (0, 1)$ for $\rho = 0$.

When $\rho \to +\infty$, the basin of attraction of $(0, 1)$ tends to invade the domain; thus the first population tends to have almost no chance of victory and the second population tends to win. However, the dependence on the parameter ρ is not monotone as one could think, at least not in $[0, +\infty) \times [0, +\infty)$.

6.2 Dependence on the Parameter ρ

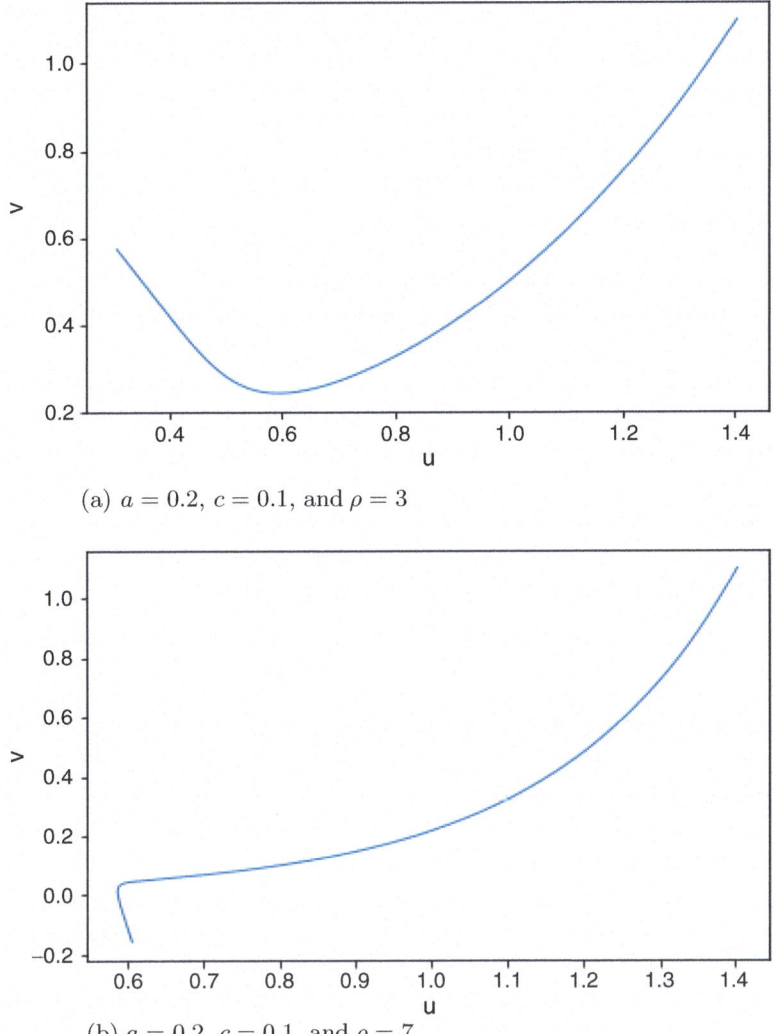

(a) $a = 0.2$, $c = 0.1$, and $\rho = 3$

(b) $a = 0.2$, $c = 0.1$, and $\rho = 7$

Fig. 6.1 (**a**) and (**b**) show the trajectory starting from the point $(u_0, v_0) = (1.4045, 1.1)$ for $\rho = 3$ and $\rho = 7$ respectively. For $\rho = 3$ the trajectory leads to the equilibrium $(0, 1)$, so $(u_0, v_0) \notin \mathcal{E}(\rho = 3)$, while for $\rho = 7$ the second population goes to extinction in finite time, so $(u_0, v_0) \in \mathcal{E}(\rho = 7)$

Indeed, by performing some simulation, one could find some values ρ_1 and ρ_2, with $0 < \rho_1 < \rho_2$, and a point $(u^*, v^*) \in [0, +\infty) \times [0, +\infty)$ such that $(u^*, v^*) \notin \mathcal{E}(\rho_1)$ and $(u^*, v^*) \in \mathcal{E}(\rho_2)$, see Fig. 6.1.

This means that, sometimes, a big value of fitness for the second population may lead to extinction, while a small value brings to victory. This is counterintuitive but can be easily explained: The parameter ρ is multiplied by the term $1 - u - v$, which is negative past the counterdiagonal of the square $[0, 1] \times [0, 1]$. So in the model (2.1), as well as in any model of Lotka–Volterra type, the population that grows faster is also the one that suffers more the consequences of overpopulation. Moreover, the usual dynamics of Lotka–Volterra models is altered by the presence of the term $-au$, and this leads to the lack of monotonicity that we observe.

We now give the proof of Proposition 6.2:

Proof of Proposition 6.2

(i) For $\rho = 0$, the equation $\dot{v} = 0$ collapses to $u = 0$. Since for $u = 0$ also the equation $\dot{u} = 0$ is satisfied, each point on the line $u = 0$ is an equilibrium.

Calculating the eigenvalues for the points $(0, \tilde{v})$, with $\tilde{v} \in [0, 1]$, using the Jacobian matrix in (5.3), one gets the values 0 and $1 - ac - \tilde{v}$. Accordingly, this entails that if $\tilde{v} < 1 - ac$, the point $(0, \tilde{v})$ corresponds to a strictly negative eigenvalue and a null one, while if $\tilde{v} > 1 - ac$, then $(0, \tilde{v})$ corresponds to a strictly negative eigenvalue and a null one. These considerations prove the first statement in (i).

We now study the behavior of the points in $[0, 1] \times [0, 1 - ac)$. Notice that the only part of its boundary that is inside $(0, 1) \times (0, 1)$ is the side $(0, 1] \times \{1 - ac\}$. We notice also that in the whole square $(0, 1] \times [0, 1]$ we have

$$\dot{v} = -au < 0,$$

so trajectories cannot exit $[0, 1] \times [0, 1 - ac)$ (see Remark 4.2).

This also gives that there is no trajectory that can go to $(0, 1)$, and there is no cycle. In particular this implies (6.11).

Thus a trajectory starting in $[0, 1] \times [0, 1-ac)$ either converges to one of the equilibria on the side $\{0\} \times [0, 1]$ or has a finite stopping time.

In particular, since $\{0\} \times [0, 1 - ac)$ consists of repulsive equilibria, we have that

$$[0, 1] \times [0, 1 - ac) \subseteq \mathcal{E}(0),$$

that is, trajectories starting in $[0, 1] \times [0, 1-ac)$ go to the extinction of v. This proves the first inclusion in (6.12).

To prove the second inclusion in (6.12), we first show that

$$\begin{aligned}&\text{points in } \big([0, 1] \times [0, 1]\big) \setminus \mathcal{T}_{\varepsilon,\delta} \text{ are mapped} \\ &\text{into } \big([0, 1] \times [0, 1]\big) \setminus \mathcal{T}_{\varepsilon,\delta} \text{ itself.}\end{aligned} \quad (6.15)$$

Indeed, on the line $\{\delta v - \varepsilon u = \delta(1 - \varepsilon)\}$, we have that the inward pointing normal derivative is given by

6.2 Dependence on the Parameter ρ

$$(\dot{u}, \dot{v}) \cdot \frac{(-\varepsilon, \delta)}{\sqrt{\varepsilon^2 + \delta^2}}$$

$$= \frac{1}{\sqrt{\varepsilon^2 + \delta^2}} (\delta\dot{v} - \varepsilon\dot{u})$$

$$= \frac{1}{\sqrt{\varepsilon^2 + \delta^2}} \big(-\delta a u - \varepsilon u(1 - u - v) + \varepsilon a c u \big) \tag{6.16}$$

$$= \frac{u}{\sqrt{\varepsilon^2 + \delta^2}} \left[\varepsilon \left(-1 + ac + u + \frac{\varepsilon}{\delta} u + 1 - \varepsilon \right) - \delta a \right]$$

$$= \frac{1}{\sqrt{\varepsilon^2 + \delta^2}} \left[u^2 \left(1 + \frac{\varepsilon}{\delta} \right) + u(\varepsilon a c - \delta a - \varepsilon^2) \right].$$

The first term is always positive, and the second one is positive for the choice

$$\delta < \frac{\varepsilon c}{2} \quad \text{and} \quad \varepsilon < \frac{ac}{2}.$$

Hence, under the assumption in (i), on the line $\{\delta v - \varepsilon u = \delta(1 - \varepsilon)\}$ the inward pointing normal derivative is positive, which implies that no trajectories in $([0, 1] \times [0, 1]) \setminus \mathcal{T}_{\varepsilon,\delta}$ can exit from $([0, 1] \times [0, 1]) \setminus \mathcal{T}_{\varepsilon,\delta}$ (see Remark 4.2). This establishes (6.15).

As a consequence of (6.15), we obtain also the second inclusion (6.12), as desired.

(ii) We claim that

$$([0, 1] \times [0, 1]) \setminus \mathcal{T}_{\varepsilon,\delta} \subseteq \mathcal{B}(\rho), \tag{6.17}$$

for all $0 < \rho < a/3$. To this end, we observe that, in order to determine the sign of the inward pointing normal derivative on the side $\{\delta v - \varepsilon u = \delta(1 - \varepsilon)\}$, by (6.16) we have to check that $\delta\dot{v} - \varepsilon\dot{u} \geqslant 0$. In order to simplify the calculation, we use the change of coordinates $x := u$ and $y := 1 - v$. In this way, one needs to verify that

$$\delta\dot{y} + \varepsilon\dot{x} < 0 \quad \text{on the line} \quad \{\delta y + \varepsilon x = \delta\varepsilon\}. \tag{6.18}$$

For this, we compute

$$\delta\dot{y} + \varepsilon\dot{x}$$
$$= \delta\rho(y - 1)(y - x) + \delta a x + \varepsilon x(y - x) - \varepsilon a c x$$
$$= -\delta\rho(1 - y)y + x\big(\delta\rho(1 - y) + \delta a + \varepsilon(y - x) - \varepsilon a c\big) \tag{6.19}$$
$$= -\delta\rho(1 - y)y + x\big(\delta\rho - \delta\rho y + \delta a + \varepsilon y - \varepsilon x - \varepsilon a c\big)$$
$$\leqslant x\big(\delta\rho - \delta\rho y + \delta a + \varepsilon y - \varepsilon x - \varepsilon a c\big).$$

Now we choose $\delta < \varepsilon c/2$, and we recall that $\rho < a/3$. Moreover, we notice that

$$y = \varepsilon - \frac{\varepsilon}{\delta} x \leqslant \varepsilon,$$

and therefore $\varepsilon y \leqslant \varepsilon^2$. Thus, we have that

$$-\delta \rho y + \delta \rho + \delta a + \varepsilon y - \varepsilon x - \varepsilon ac \leqslant \frac{\varepsilon ac}{6} + \frac{\varepsilon ac}{2} + \varepsilon^2 - \varepsilon ac$$

$$= \varepsilon \left(\frac{2}{3} ac + \varepsilon - ac \right),$$

which is negative for $\varepsilon < ac/3$. Plugging this information into (6.19), we obtain (6.18), as desired.

This proves that trajectories in $\big([0, 1] \times [0, 1)\big) \setminus \mathcal{T}_{\varepsilon,\delta}$ cannot exit $\big([0, 1] \times [0, 1]\big) \setminus \mathcal{T}_{\varepsilon,\delta}$ (see Remark 4.2). This, the fact that there are no cycles in $[0, 1] \times [0, 1]$ and the Poincaré–Bendixson Theorem (see, e.g., [90]) give that trajectories in $\big([0, 1] \times [0, 1]\big) \setminus \mathcal{T}_{\varepsilon,\delta}$ converge to $(0, 1)$, that is the only equilibrium in $\big([0, 1] \times [0, 1]\big) \setminus \mathcal{T}_{\varepsilon,\delta}$. Hence, (6.17) is established.

From (6.17) we deduce that

$$\mathcal{E}(\rho) \subseteq \mathcal{T}_{\varepsilon,\delta}$$

for all $0 < \rho < a/3$, which implies the desired result in (ii).

(iii) We consider $\varepsilon_1 > \varepsilon_2 > 0$ to be taken sufficiently small in what follows, and we show that there exists $R > 0$, depending on ε_1 and ε_2, such that for all $\rho \geqslant R$ it holds that

$$\mathcal{R}_{\varepsilon_1,\varepsilon_2} := [0, 1 - \varepsilon_1] \times [\varepsilon_2, 1] \subseteq \mathcal{B}(\rho). \tag{6.20}$$

For this, we first observe that

$$\text{no trajectory starting in } \mathcal{R}_{\varepsilon_1,\varepsilon_2} \text{ can exit the set.} \tag{6.21}$$

Indeed, looking at the velocity fields on the side $\{1 - \varepsilon_1\} \times [\varepsilon_2, 1]$, the normal inward derivative is

$$-\dot{u} = -[u(1 - u - v) - acu] = -(1 - \varepsilon_1)(\varepsilon_1 - v - ac),$$

and this is positive for $\varepsilon_1 \leqslant ac$ (which is fixed from now on). In addition, on the side $[0, 1 - \varepsilon_1] \times \{\varepsilon_2\}$, the inward normal derivative is

$$\dot{v} = [\rho v(1 - u - v) - au]$$

$$= \rho\varepsilon_2(1 - u - \varepsilon_2) - au$$
$$\geqslant \rho\varepsilon_2(\varepsilon_1 - \varepsilon_2) - a(1 - \varepsilon_1),$$

and this is positive for

$$\rho > \frac{a(1 - \varepsilon_1)}{\varepsilon_2(\varepsilon_1 - \varepsilon_2)} =: R. \tag{6.22}$$

These observations complete the proof of (6.21) (see Remark 4.2).

From (5.41), (6.21), and the Poincaré–Bendixson Theorem (see, e.g., [90]), we have that all the trajectories in the interior of $\mathcal{R}_{\varepsilon_1,\varepsilon_2}$ must converge to either an equilibrium or a union of (finitely many) equilibria and non-closed orbits connecting these equilibria.

In addition, we claim that if $0 < ac < 1$, recalling (3.5) and possibly enlarging ρ in (6.22),

$$(u_s, v_s) \notin \mathcal{R}_{\varepsilon_1,\varepsilon_2}. \tag{6.23}$$

Indeed, we have that $u_s \to 1 - ac$ and $v_s \to 0$, as $\rho \to +\infty$. Hence, we can choose ρ large enough such that the statement in (6.23) is satisfied.

As a consequence of (6.23), we get that all the trajectories in the interior of $\mathcal{R}_{\varepsilon_1,\varepsilon_2}$ must converge to the equilibrium $(0, 1)$, and this establishes (6.20).

Accordingly, (6.20) entails that, for $\varepsilon_1 > \varepsilon_2 > 0$ sufficiently small, there exists $R > 0$, depending on ε_1 and ε_2, such that for all $\rho \geqslant R$

$$\mathcal{E}(\rho) \subset \big((0, 1] \times [0, \varepsilon_2)\big) \cup \big((1 - \varepsilon_1, 1] \times (\varepsilon_2, 1]\big).$$

This implies (6.14), as desired. □

6.3 Dependence on the Parameter a

The consequences of the lack of variational structure become even more extreme when we observe the dependence of the dynamics on the parameter a, that is, the aggressiveness of the first population toward the other. Throughout this chapter, we take $\rho > 0$ and $c > 0$, and we perform our analysis taking into account the limit cases $a \to 0$ and $a \to +\infty$. We start analyzing the dynamics of (2.1) in the case $a = 0$.

Proposition 6.3 (Dynamics of (2.1) **when** $a = 0$**)** *For $a = 0$ the system* (2.1) *has the following features:*

(i) The system has the equilibrium $(0, 0)$, which is a source, and a straight line of equilibria $(u, 1 - u)$, for all $u \in [0, 1]$, which correspond to a strictly negative eigenvalue and a null one.

(ii) Given any $(u(0), v(0)) \in (0, 1) \times (0, 1)$, we have that

$$(u(t), v(t)) \to (\bar{u}, 1 - \bar{u}) \quad \text{as } t \to +\infty, \tag{6.24}$$

where $\bar{u} \in (0, 1)$ satisfies

$$\frac{v(0)}{u^\rho(0)} \bar{u}^\rho + \bar{u} - 1 = 0. \tag{6.25}$$

(iii) The equilibrium (u_s^0, v_s^0) given in (3.8) has a stable manifold, which can be written as the graph of an increasing smooth function

$$\gamma_0 : [0, u_{\mathcal{M}}^0] \to [0, v_{\mathcal{M}}^0],$$

as given in (3.11), for some

$$(u_{\mathcal{M}}^0, v_{\mathcal{M}}^0) \in \big(\{1\} \times [0, 1]\big) \cup \big((0, 1] \times \{1\}\big),$$

such that $\gamma_0(0) = 0$ and $\gamma_0(u_{\mathcal{M}}^0) = v_{\mathcal{M}}^0$.
More precisely,

$$u_{\mathcal{M}}^0 := \min\left\{1, \frac{u_s^0}{(v_s^0)^{\frac{1}{\rho}}}\right\}, \tag{6.26}$$

being (u_s^0, v_s^0) defined in (3.8).

We point out that formula (6.24) says that for $a = 0$ every point in the interior of $[0, 1] \times [0, 1]$ tends to a coexistence equilibrium. The shape of the trajectories depends on ρ, being convex in the case $\rho > 1$, a straight line in the case $\rho = 1$, and concave in the case $\rho < 1$. This means that if the second population v is alive at the initial time, then it does not get extinct in finite time.

Proof of Proposition 6.3

(i) For $a = 0$, we look for the equilibria of the system (2.1) by studying when $\dot{u} = 0$ and $\dot{v} = 0$. It is easy to see that the point $(0, 0)$ and all the points on the line $u + v = 1$ are the only equilibria.
The Jacobian of the system (see (5.3), with $a = 0$) at the point $(0, 0)$ has two positive eigenvalues, 1 and ρ, and therefore $(0, 0)$ is a source.

6.3 Dependence on the Parameter a

Furthermore, the characteristic polynomial at a point $(\tilde{u}, \tilde{v})$ on the line $u + v = 1$ is given by

$$(\lambda + \tilde{u})(\lambda + \rho\tilde{v}) - \rho\tilde{u}\tilde{v} = \lambda(\lambda + \tilde{u} + \rho\tilde{v}),$$

and therefore, the eigenvalues are 0 and $-\tilde{u} - \rho\tilde{v} < 0$.

(ii) We point out that when $a = 0$

$$\mu(t) := v(t)/u^\rho(t) \text{ is a prime integral for the system.} \tag{6.27}$$

Indeed,

$$\dot{\mu} = \frac{\dot{v}u^\rho - \rho u^{\rho-1}\dot{u}v}{u^{2\rho}}$$

$$= u^{\rho-1}\frac{\rho uv(1 - u - v) - \rho uv(1 - u - v)}{u^{2\rho}}$$

$$= 0.$$

As a result, the trajectory starting at a point $(u(0), v(0)) \in (0, 1) \times (0, 1)$ lies on the curve

$$v(t) = \frac{v(0)}{u^\rho(0)} u^\rho(t). \tag{6.28}$$

Notice that these curves do not intersect the line $\{v = 0\}$ and that they are not periodic orbits, since $\dot{u}$ and $\dot{v}$ have the same sign. Hence, the α-limit point of $(u(0), v(0))$ is an equilibrium on this curve. Since $(0, 0)$ is a source, the only possibility is that the trajectory starting at $(u(0), v(0))$ converges to an equilibrium $(\bar{u}, \bar{v})$ such that $\bar{v} = 1 - \bar{u}$. This entails that

$$1 - \bar{u} = \bar{v} = (v(0)/u^\rho(0))\bar{u}^\rho,$$

which is exactly Eq. (6.25).

(iii) We observe that the point (u_s^0, v_s^0) given in (3.8) lies on the straight line $u + v = 1$, and therefore, thanks to (i) here, it is an equilibrium of the system (2.1), which corresponds to a strictly negative eigenvalue $-u_s^0 - \rho v_s^0$ and a null one.

Hence, by the Center Manifold Theorem (see, e.g., Theorem 1 on page 16 of [18]), the point (u_s^0, v_s^0) has a stable manifold, which has dimension 1 and is tangent to the eigenvector of the linearized system associated with the strictly negative eigenvalue $-u_s^0 - \rho v_s^0$.

Also, the monotonicity and the property of being a graph follow from the strict sign of $\dot u$ and $\dot v$. The smoothness of the graphs follows from the regularity of the center manifold. The fact that $\gamma_0(0) = 0$ is a consequence of the monotonicity property of u and v, which ensures that the ω-limit exists and the fact that this limit has to lie on the prime integral in (6.28). The fact that $\gamma_0(u_\mathcal{M}^0) = v_\mathcal{M}^0$ follows from formula (6.24) and the monotonicity property. Formula (3.11) follows from the fact that any trajectory has to lie on the prime integral in (6.28). $\square$

To state our next result concerning the dependence of $\mathcal{E}$ defined in (3.4) on the parameter a, we give some notation. We will make it explicit the dependence of the sets $\mathcal{E}$ and $\mathcal{B}$ on the parameter a, by writing explicitly $\mathcal{E}(a)$ and $\mathcal{B}(a)$, and we will call

$$\mathcal{E}_0 := \bigcap_{a'>0} \bigcup_{a'>a>0} \mathcal{E}(a)$$

and

$$\mathcal{E}_\infty := \bigcap_{a'>0} \bigcup_{a>a'} \mathcal{E}(a). \tag{6.29}$$

In this setting, we have the following statements:

Proposition 6.4 (Dependence of the Dynamics On a)

(i) We have that

$$\mathcal{G} \subseteq \mathcal{E}_0 \subseteq \overline{\mathcal{G}}, \tag{6.30}$$

where

$$\mathcal{G} := \big\{(u, v) \in [0, 1] \times [0, 1] \ \text{s.t.} \ v < \gamma_0(u) \ \text{if} \ u \in [0, u_\mathcal{M}^0]$$
$$\text{and} \ v \leqslant 1 \ \text{if} \ u \in (u_\mathcal{M}^0, 1]\big\},$$

and γ_0 and $u_\mathcal{M}^0$ are given in (3.11).
(ii) It holds that

$$\mathcal{S}_c \subseteq \mathcal{E}_\infty \subseteq \overline{\mathcal{S}_c}, \tag{6.31}$$

where

$$\mathcal{S}_c := \Big\{(u, v) \in [0, 1] \times [0, 1] \ \text{s.t.} \ v - \frac{u}{c} < 0\Big\}. \tag{6.32}$$

6.3 Dependence on the Parameter a

We point out that the set $\mathcal{E}_0$ in (6.30) does not coincide with the basin of attraction for the system (2.1) when $a = 0$. Indeed, as already mentioned, formula (6.24) in Proposition 6.3 says that for $a = 0$ every point in the interior of $[0, 1] \times [0, 1]$ tends to a coexistence equilibrium, and thus if $v(0) \neq 0$, then $v(t)$ does not get extinct in finite time.

Also, as $a \to +\infty$, we have that the set $\mathcal{E}_\infty$ is determined by $\mathcal{S}_c$, defined in (6.32), that depends only on the parameter c.

The statement in (i) of Proposition 6.4 will be a direct consequence of the following result. Recalling the function γ introduced in Propositions 5.1 and 5.7, we express here the dependence on the parameter a by writing γ_a, u_a, v_a, u_s^a, $u_{\mathcal{M}}^a$. We will also denote by $\mathcal{M}^a$ the stable manifold of the point (u_s, v_s) in (3.5) and by $\mathcal{M}^0$ the stable manifold of the point (u_s^0, v_s^0) in (3.8). The key lemma is the following:

Lemma 6.5 *For all $u \in [0, 1]$, we have that $\gamma_a(u) \to \gamma_0(u)$ uniformly as $a \to 0$, where $\gamma_0(u)$ is the function defined in (3.11).*

Proof Since we are dealing with the limit as a goes to zero, throughout this proof we will always assume that we are in the case $ac < 1$.

Also, we denote by $\phi_p^a(t)$ the flow at time t of the point $p \in [0, 1] \times [0, 1]$ associated with (2.1) and similarly by $\phi_p^{(0)}(t)$ the flow at time t of the point p associated with (2.1) when $a = 0$. With a slight abuse of notation, we will also write

$$\phi_p^a(t) = (u_a(t), v_a(t)) \quad \text{with} \quad p = (u_a(0), v_a(0)).$$

Let us start by proving that

$$\mathcal{M}^a \cap \left([0, u_s^0] \times [0, v_s^0]\right) \to \mathcal{M}^0 \cap \left([0, u_s^0] \times [0, v_s^0]\right) \quad \text{as } a \to 0. \tag{6.33}$$

For this, we claim that, for every $\varepsilon > 0$, if

$$(u_a(0))^2 + (v_a(0))^2 \geq \frac{\varepsilon^2}{4} \tag{6.34}$$

and

$$\left|(u_a(t), v_a(t)) - (u_s^a, v_s^a)\right| > \frac{\varepsilon}{2}, \tag{6.35}$$

then

$$|\dot{u}_a(t)|^2 + |\dot{v}_a(t)|^2 > \frac{\varepsilon^4}{C_0}, \tag{6.36}$$

for some $C_0 > 0$, depending only on ρ and c.

Indeed, by recalling (v) of Theorem 3.2 and (6.35), we see that the trajectory $(u_a(t), v_a(t))$ belongs to the set

$$[0, u_s^a] \times [0, v_s^a] \setminus B_{\frac{\varepsilon}{2}}(u_s^a, v_s^a).$$

Moreover, we claim that

$$1 - ac - u_a(t) - v_a(t) \geqslant \frac{\varepsilon\sqrt{2}}{4}, \qquad (6.37)$$

for any $t > 0$ such that (6.35) is satisfied. To prove this, we recall that (u_s^a, v_s^a) lies on the straight line ℓ given by $v = -u + 1 - ac$ when $0 < ac < 1$ (see (5.1)). Clearly, there is no point of the set

$$[0, u_s^a] \times [0, v_s^a] \setminus B_{\frac{\varepsilon}{2}}(u_s^a, v_s^a)$$

lying on ℓ, and we notice that the points in the set

$$[0, u_s^a] \times [0, v_s^a] \setminus B_{\frac{\varepsilon}{2}}(u_s^a, v_s^a)$$

with minimal distance from ℓ are given by

$$p := \left(u_s^a - \frac{\varepsilon}{2}, v_s^a\right) \qquad \text{and} \qquad q := \left(u_s^a, v_s^a - \frac{\varepsilon}{2}\right).$$

Also, the distance of the point p from the straight line ℓ is given by

$$\frac{\varepsilon}{2} \cdot \tan\frac{\pi}{4} = \frac{\varepsilon\sqrt{2}}{4}.$$

Thus, the distance between $(u_a(t), v_a(t))$ and the line ℓ is greater than

$$\frac{\varepsilon\sqrt{2}}{4},$$

and this gives (6.37).

As a consequence of (6.37), we obtain that

$$(\dot{u}_a(t))^2 = \left(u_a(t)(1 - ac - u_a(t) - v_a(t))\right)^2$$
$$> (u_a(t))^2 \left(\frac{\varepsilon\sqrt{2}}{4}\right)^2 \qquad (6.38)$$

6.3 Dependence on the Parameter a

and that

$$\begin{aligned}(\dot{v}_a(t))^2 &= \left(\rho v_a(t)(1 - u_a(t) - v_a(t)) - au_a(t)\right)^2 \\ &\geqslant \left(\rho v_a(t)\left(ac + \frac{\varepsilon\sqrt{2}}{4}\right) - au_a(t)\right)^2.\end{aligned} \quad (6.39)$$

Now, if $u_a(t) \geqslant \rho c v_a(t)$, then from (6.38) and (6.34) we obtain that

$$\begin{aligned}(\dot{u}_a(t))^2 + (\dot{v}_a(t))^2 &\geqslant (\dot{u}_a(t))^2 \\ &> (u_a(t))^2 \left(\frac{\varepsilon\sqrt{2}}{4}\right)^2 \\ &\geqslant \frac{(u_a(t))^2}{2}\left(\frac{\varepsilon\sqrt{2}}{4}\right)^2 + \frac{(\rho c v_a(t))^2}{2}\left(\frac{\varepsilon\sqrt{2}}{4}\right)^2 \\ &\geqslant \min\{1, \rho^2 c^2\}\frac{\varepsilon^2}{16}\left((u_a(t))^2 + (v_a(t))^2\right) \\ &\geqslant \min\{1, \rho^2 c^2\}\frac{\varepsilon^2}{16}\left((u_a(0))^2 + (v_a(0))^2\right) \\ &\geqslant \min\{1, \rho^2 c^2\}\frac{\varepsilon^4}{64},\end{aligned}$$

which proves (6.36) in this case.

If instead $u_a(t) < \rho c v_a(t)$, we use (6.39) to see that

$$\begin{aligned}(\dot{u}_a(t))^2 + (\dot{v}_a(t))^2 &\geqslant (\dot{v}_a(t))^2 \\ &\geqslant \left(\rho v_a(t)\left(ac + \frac{\varepsilon\sqrt{2}}{4}\right) - au_a(t)\right)^2 \\ &= \left(\frac{\varepsilon\sqrt{2}\rho v_a(t)}{4} + a\left(\rho c v_a(t) - u_a(t)\right)\right)^2 \\ &\geqslant \left(\frac{\varepsilon\sqrt{2}\rho v_a(t)}{4}\right)^2 \\ &\geqslant \frac{1}{2}\left(\frac{\varepsilon\sqrt{2}\rho v_a(t)}{4}\right)^2 + \frac{1}{2}\left(\frac{\varepsilon\sqrt{2}u_a(t)}{4c}\right)^2 \\ &\geqslant \min\left\{\rho^2, \frac{1}{c^2}\right\}\frac{\varepsilon^2}{16}\left((u_a(t))^2 + (v_a(t))^2\right)\end{aligned}$$

$$\geqslant \min\left\{\rho^2, \frac{1}{c^2}\right\} \frac{\varepsilon^2}{16}\left((u_a(0))^2 + (v_a(0))^2\right)$$

$$\geqslant \min\left\{\rho^2, \frac{1}{c^2}\right\} \frac{\varepsilon^4}{64},$$

which completes the proof of (6.36).

Now, for any $\eta > 0$, we define

$$\mathcal{P}_\eta := \left\{(u, v) \in [0, 1] \times [0, 1] \text{ s.t.} \right.$$

$$\left. v = \frac{v_s^0 - \eta'}{(u_s^0 + \eta')^\rho} u^\rho \text{ with } |\eta'| \leqslant \eta \right\}.$$

Notice that this is the union of graphs of functions close to the curve γ_0. Given $\varepsilon > 0$, we define

$$\eta(\varepsilon) \text{ to be the smallest } \eta \text{ for which } \mathcal{P}_\eta \supset B_\varepsilon(u_s^0, v_s^0). \tag{6.40}$$

We remark that

$$\lim_{\varepsilon \to 0} \eta(\varepsilon) = 0. \tag{6.41}$$

Also, given $\delta > 0$, we define a tubular neighborhood $\mathcal{U}_\delta$ of $\mathcal{M}^0$ as

$$\mathcal{U}_\delta := \bigcup_{q \in \mathcal{M}^0} B_\delta(q).$$

Furthermore, we define

$$\delta(\varepsilon) \text{ the smallest } \delta \text{ such that } \mathcal{U}_\delta \supset \mathcal{P}_{\eta(\varepsilon)}. \tag{6.42}$$

Recalling (6.41), we have that

$$\lim_{\varepsilon \to 0} \delta(\varepsilon) = 0. \tag{6.43}$$

We remark that, as $a \to 0$, the point (u_s^a, v_s^a) in (3.5), which is a saddle point for the dynamics of (2.1) when $ac < 1$ (recall Theorem 3.2), tends to the point (u_s^0, v_s^0) in (3.8), which belongs to the line $v + u = 1$, which is an equilibrium point for the dynamics of (2.1) when $a = 0$, according to Proposition 6.3.

6.3 Dependence on the Parameter a

As a consequence, for every $\varepsilon > 0$, there exists $a_\varepsilon > 0$ such that if $a \in (0, a_\varepsilon)$,

$$|(u_s^a, v_s^a) - (u_s^0, v_s^0)| \leq \frac{\varepsilon}{8}. \tag{6.44}$$

This gives that the intersection of $\mathcal{M}^a$ with $B_{\varepsilon/2}(u_s^0, v_s^0)$ is nonempty.

Furthermore, since $\gamma_a(0) = 0$, in light of Proposition 5.1, we have that the intersection of $\mathcal{M}^a$ with $B_{\varepsilon/2}$ is nonempty. Hence, there exists $p_{\varepsilon,a} \in \mathcal{M}^a \cap \partial B_{\varepsilon/2}$.

We also notice that

$$\mathcal{M}^a = \phi^a_{p_{\varepsilon,a}}(\mathbb{R}). \tag{6.45}$$

In addition,

$$\phi^a_{p_{\varepsilon,a}}\big((-\infty, 0]\big) \subset B_{\varepsilon/2}. \tag{6.46}$$

Also, since the origin belongs to $\mathcal{M}^0$, we have that $B_{\varepsilon/2} \subset \mathcal{U}_\varepsilon$. From this and (6.46), we deduce that

$$\phi^a_{p_{\varepsilon,a}}\big((-\infty, 0]\big) \subset \mathcal{U}_\varepsilon. \tag{6.47}$$

Now, we let C_0 be as in (6.36), and we claim that there exists $t_{\varepsilon,a} \in (0, 3\sqrt{C_0}\varepsilon^{-2})$ such that

$$\phi^a_{p_{\varepsilon,a}}(t_{\varepsilon,a}) \in \partial B_{3\varepsilon/4}(u_s^0, v_s^0). \tag{6.48}$$

To check this, we argue by contradiction, and we suppose that

$$\phi^a_{p_{\varepsilon,a}}\big((0, 3\sqrt{C_0}\varepsilon^{-2})\big) \cap B_{3\varepsilon/4}(u_s^0, v_s^0) = \varnothing.$$

Then, for every $t \in (0, 3\sqrt{C_0}\varepsilon^{-2})$, recalling also (6.44),

$$\big|\phi^a_{p_{\varepsilon,a}}(t) - (u_s^a, v_s^a)\big| \geq \big|\phi^a_{p_{\varepsilon,a}}(t) - (u_s^0, v_s^0)\big| - \big|(u_s^a, v_s^a) - (u_s^0, v_s^0)\big|$$

$$\geq \frac{3\varepsilon}{4} - \frac{\varepsilon}{8}$$

$$> \frac{\varepsilon}{2},$$

and consequently (6.35) is satisfied for every $t \in (0, 3\sqrt{C_0}\varepsilon^{-2})$.

Moreover, we observe that $p_{\varepsilon,a}$ satisfies (6.34), and therefore, by (6.36),

$$|\dot{u}_a(t)|^2 + |\dot{v}_a(t)|^2 > \frac{\varepsilon^4}{C_0},$$

for all $t \in (0, 3\sqrt{C_0}\varepsilon^{-2})$, where we used the notation

$$\phi_{p_{\varepsilon,a}}^a(t) = (u_a(t), v_a(t)),$$

being

$$p_{\varepsilon,a} = (u_a(0), v_a(0)).$$

As a result,

$$\left(\dot{u}_a(t) + \dot{v}_a(t)\right)^2 > \frac{\varepsilon^4}{C_0},$$

and thus

$$\dot{u}_a(t) + \dot{v}_a(t) > \frac{\varepsilon^2}{\sqrt{C_0}}.$$

This leads to

$$u_a\left(\frac{3\sqrt{C_0}}{\varepsilon^2}\right) + v_a\left(\frac{3\sqrt{C_0}}{\varepsilon^2}\right)$$

$$= u_a(0) + v_a(0) + \int_0^{\frac{3\sqrt{C_0}}{\varepsilon^2}} \left(\dot{u}_a(t) + \dot{v}_a(t)\right) dt$$

$$\geq u_a(0) + v_a(0) + \int_0^{\frac{3\sqrt{C_0}}{\varepsilon^2}} \frac{\varepsilon^2}{\sqrt{C_0}} dt$$

$$= u_a(0) + v_a(0) + 3$$

$$\geq 3,$$

which forces the trajectory to exit the region $[0, 1] \times [0, 1]$. This is against the assumption that $p_{\varepsilon,a} \in \mathcal{M}^a$, and therefore the proof of (6.48) is complete.

In light of (6.48), we can set $q_{\varepsilon,a} := \phi_{p_{\varepsilon,a}}^a(t_{\varepsilon,a})$, and we deduce from (6.40) that $q_{\varepsilon,a} \in \mathcal{P}_{\eta(\varepsilon)}$. We also observe that the set $\mathcal{P}_\eta$ is invariant for the flow with $a = 0$, thanks to (6.27). These observations give that $\phi_{q_{\varepsilon,a}}^0(t) \in \mathcal{P}_{\eta(\varepsilon)}$ for all $t \in \mathbb{R}$.

6.3 Dependence on the Parameter a

As a result, using (6.42), we conclude that

$$\phi^0_{q_{\varepsilon,a}}(t) \in \mathcal{U}_{\delta(\varepsilon)} \quad \text{for all } t \in \mathbb{R}. \tag{6.49}$$

In addition, by the continuous dependence of the flow on the parameter a in closed intervals of time (see, e.g., Section 2.4 in [37] or Theorem 2.4.2 in [49])

$$\left|\phi^0_{q_{\varepsilon,a}}(t) - \phi^a_{q_{\varepsilon,a}}(t)\right| < \varepsilon,$$

for all $t \in [-3\sqrt{C_0}\varepsilon^{-2}, 0]$, provided that a is sufficiently small, possibly in dependence of ε. This fact and (6.49) entail that

$$\phi^a_{q_{\varepsilon,a}}(t) \in \mathcal{U}_{\delta(\varepsilon)+\varepsilon} \quad \text{for all } t \in [-3\sqrt{C_0}\varepsilon^{-2}, 0].$$

In particular, for all $t \in [0, t_{\varepsilon,a}]$,

$$\phi^a_{p_{\varepsilon,a}}(t) = \phi^a_{q_{\varepsilon,a}}(t - t_{\varepsilon,a}) \in \mathcal{U}_{\delta(\varepsilon)+\varepsilon}. \tag{6.50}$$

We now claim that for all $t \geq t_{\varepsilon,a}$,

$$\phi^a_{p_{\varepsilon,a}}(t) \subset B_\varepsilon(u^a_s, v^a_s). \tag{6.51}$$

Indeed, this is true when $t = t_{\varepsilon,a}$, thanks to (6.44) and (6.48).

Hence, since the trajectory $\phi^a_{p_{\varepsilon,a}}(t)$ is contained in the domain where $\dot{u} \geq 0$ and $\dot{v} \geq 0$, thanks to (5.6), we deduce that (6.51) holds true.

From (6.44) and (6.51), we conclude that

$$\phi^a_{p_{\varepsilon,a}}(t) \subset B_{2\varepsilon}(u^0_s, v^0_s),$$

for all $t \geq t_{\varepsilon,a}$.

Using this, (6.47), and (6.50), we obtain that

$$\phi^a_{p_{\varepsilon,a}}(\mathbb{R}) \subset \mathcal{U}_{\delta(\varepsilon)+2\varepsilon}.$$

This and (6.43) give that (6.33) is satisfied, as desired.

One can also show that

$$\mathcal{M}^a \cap \left([u^0_s, u^0_\mathcal{M}] \times [v^0_s, v^0_\mathcal{M}]\right) \to \mathcal{M}^0 \cap \left([u^0_s, u^0_\mathcal{M}] \times [v^0_s, v^0_\mathcal{M}]\right) \tag{6.52}$$

as $a \to 0$. The proof of (6.52) is similar to that of (6.33), just replacing $p_{\varepsilon,a}$ with $(u^a_\mathcal{M}, v^a_\mathcal{M})$ (in this case the analysis near the origin is simply omitted since the trajectory has only one limit point).

With (6.33) and (6.52), the proof of Lemma 6.5 is thereby complete. □

Now we are ready to give the proof of Proposition 6.4:

Proof of Proposition 6.4

(i) We aim at proving that $\mathcal{G} \subseteq \mathcal{E}_0 \subseteq \overline{\mathcal{G}}$.

For this, we observe that, by Lemma 6.5, $\gamma_a(u)$ converges to $\gamma_0(u)$ pointwise as $a \to 0$. In particular, $u_\mathcal{M}^a \to u_\mathcal{M}^0$ as $a \to 0$.

Also, recalling (3.11), we notice that if $u_\mathcal{M}^0 = u_s^0/(v_s^0)^{\frac{1}{\rho}} < 1$, then $\gamma_0(u_\mathcal{M}^0) = 1$; otherwise if $u_\mathcal{M}^0 = 1$, then $\gamma_0(u_\mathcal{M}^0) < 1$, being $\gamma_0(u)$ strictly monotone increasing. Furthermore, thanks to Proposition 5.9, we know that the set $\mathcal{E}(a)$ is bounded from above by the graph of the function $\gamma_a(u)$ for $u \in [0, u_\mathcal{M}^a]$ and from the straight line $v = 1$ for $u \in (u_\mathcal{M}^a, 1]$ (i.e., nonempty for $u_\mathcal{M}^a < 1$).

Now we claim that, for all $a' > 0$,

$$\mathcal{G} \subseteq \bigcup_{0 < a < a'} \mathcal{E}(a). \tag{6.53}$$

To show this, we take a point $(u, v) \in \mathcal{G}$. Hence, in light of the considerations above, we have that $(u, v) \in \mathcal{E}(a)$ for any a sufficiently small, which proves (6.53).

From (6.53), we deduce that

$$\mathcal{G} \subseteq \bigcap_{a' > 0} \bigcup_{0 < a < a'} \mathcal{E}(a). \tag{6.54}$$

Now we show that

$$\bigcap_{a' > 0} \bigcup_{0 < a < a'} \mathcal{E}(a) \subseteq \overline{\mathcal{G}}. \tag{6.55}$$

For this, we take

$$(\widehat{u}, \widehat{v}) \in \bigcap_{a' > 0} \bigcup_{0 < a < a'} \mathcal{E}(a),$$

and then it must hold that for every $a' > 0$ there exists $a < a'$ such that $(\widehat{u}, \widehat{v}) \in \mathcal{E}(a)$, namely $\widehat{v} < \gamma_a(\widehat{u})$ if $\widehat{u} \in [0, u_\mathcal{M}^a]$ and $\widehat{v} \leq 1$ if $\widehat{u} \in (u_\mathcal{M}^a, 1]$. Thus, by the pointwise convergence, we have that $\widehat{v} \leq \gamma_0(\widehat{u})$ if $\widehat{u} \in [0, u_\mathcal{M}^0]$ and $\widehat{v} \leq 1$ if $\widehat{u} \in (u_\mathcal{M}^0, 1]$, which proves (6.55).

From (6.54) and (6.55), we conclude that

$$\mathcal{G} \subseteq \bigcap_{a' > 0} \bigcup_{0 < a < a'} \mathcal{E}(a) = \mathcal{E}_0 \subseteq \overline{\mathcal{G}},$$

6.3 Dependence on the Parameter a

as desired.

(ii) Since we deal with the limit case as $a \to +\infty$, we suppose from now on that $ac > 1$. We fix $\varepsilon > 0$, and we consider the set

$$\mathcal{S}_{\varepsilon^+} := \left\{ (u, v) \in [0, 1] \times [0, 1] \text{ s.t. } v > u\left(\frac{1}{c} + \varepsilon\right) \right\}.$$

We claim that

$$\mathcal{S}_{\varepsilon^+} \subseteq \mathcal{B}(a) \tag{6.56}$$

for a big enough, possibly in dependence of ε. For this, we first analyze the component of the velocity in the inward normal directions along the boundary of $\mathcal{S}_{\varepsilon^+}$. The only side on the interior of $[0, 1] \times [0, 1]$ is given by the straight line $v - u(\varepsilon + 1/c) = 0$. Ignoring the scaling constant $1/\sqrt{1 + \varepsilon^2 + \frac{2\varepsilon}{c} + \frac{1}{c^2}}$, we compute

$$(\dot u, \dot v) \cdot \left(-\left(\varepsilon + \frac{1}{c}\right), 1 \right) = \dot v - \dot u \left(\varepsilon + \frac{1}{c}\right)$$

$$= \rho v(1 - u - v) - au - \left(\varepsilon + \frac{1}{c}\right) u(1 - u - v) + \left(\varepsilon + \frac{1}{c}\right) acu$$

$$= \left[\rho v - \left(\varepsilon + \frac{1}{c}\right) u \right] (1 - u - v) + \varepsilon a c u.$$

Thus, by using that $v - u(\varepsilon + 1/c) = 0$, we obtain that

$$(\dot u, \dot v) \cdot \left(-\left(\varepsilon + \frac{1}{c}\right), 1 \right) = u \left[a\varepsilon c + (\rho - 1)(1 - u - v)\left(\varepsilon + \frac{1}{c}\right) \right].$$

Notice that $u \leqslant 1$ and $|1 - u - v| \leqslant 2$, and therefore

$$(\dot u, \dot v) \cdot \left(-\left(\varepsilon + \frac{1}{c}\right), 1 \right) \geqslant u \left[a\varepsilon c - 2(\rho + 1)\left(\varepsilon + \frac{1}{c}\right) \right].$$

Accordingly, the normal velocity is positive for $a \geqslant a_1$, where

$$a_1 := 2(\rho + 1)\left(\varepsilon + \frac{1}{c}\right)\frac{1}{\varepsilon c}.$$

Hence, by Lemma 4.7, no trajectory can exit $\mathcal{S}_{\varepsilon^+}$. These considerations, together with the fact that there are no cycles in $[0, 1] \times [0, 1]$ and that $\mathcal{S}_{\varepsilon^+} \cap \{v = 0\} = \emptyset$ and the Poincaré–Bendixson Theorem (see, e.g., [90]), give that the ω-limit set of any trajectory starting in the interior of $\mathcal{S}_{\varepsilon^+}$ can be either an equilibrium or a union of (finitely many) equilibria and non-closed orbits connecting these equilibria.

We remark that

$$\text{the } \omega\text{-limit set of any trajectory cannot be the equilibrium } (0, 0). \tag{6.57}$$

Indeed, if the ω-limit of a trajectory were $(0, 0)$, then this trajectory would lie on the stable manifold of $(0, 0)$, and moreover it must be contained in $\mathcal{S}_{\varepsilon^+}$, since no trajectory can exit $\mathcal{S}_{\varepsilon^+}$. On the other hand, by Proposition 5.1, we have that at $u = 0$ the stable manifold is tangent to the line

$$v = \frac{a}{\rho - 1 + ac} u = \frac{1}{\frac{\rho-1}{a} + c} u.$$

Now, if we take a sufficiently large, this line lies below the line $v = u(1/c + \varepsilon)$, thus providing a contradiction. Hence, the proof of (6.57) is complete.

Accordingly, since $(0, 1)$ is a sink, the only possibility is that the ω-limit set of any trajectory starting in the interior of $\mathcal{S}_{\varepsilon^+}$ is the equilibrium $(0, 1)$. Namely, we have established (6.56).

As a consequence of (6.56), we deduce that for every $\varepsilon > 0$ there exists $a_\varepsilon > 0$ such that

$$\bigcup_{a \geqslant a_\varepsilon} \mathcal{E}(a) \subseteq \left\{ (u, v) \in [0, 1] \times [0, 1] \text{ s.t. } v \leqslant u\left(\frac{1}{c} + \varepsilon\right) \right\}. \tag{6.58}$$

In addition,

$$\bigcap_{\varepsilon > 0} \left\{ (u, v) \in [0, 1] \times [0, 1] \text{ s.t. } v \leqslant u\left(\frac{1}{c} + \varepsilon\right) \right\}$$
$$= \left\{ (u, v) \in [0, 1] \times [0, 1] \text{ s.t. } v \leqslant \frac{u}{c} \right\} = \overline{\mathcal{S}_c}.$$

This and (6.58) entail that

$$\bigcap_{a' > 0} \bigcup_{a > a'} \mathcal{E}(a) \subseteq \overline{\mathcal{S}_c},$$

which implies the second inclusion in (6.31).

Now, to show the first inclusion in (6.31), for every $\varepsilon \in (0, 1/c)$ we consider the set

$$\mathcal{S}_{\varepsilon^-} := \left\{ (u, v) \in [0, 1] \times [0, 1] \text{ s.t. } v < u\left(\frac{1}{c} - \varepsilon\right) \right\}.$$

We claim that, for all $\varepsilon \in (0, 1/c)$,

6.3 Dependence on the Parameter a

$$\mathcal{S}_{\varepsilon^-} \subseteq \mathcal{E}_\infty. \tag{6.59}$$

For this, we first show that if a is sufficiently large, possibly in dependence of ε,

$$\mathcal{S}_{\varepsilon^-} \subseteq \mathcal{E}(a). \tag{6.60}$$

Indeed, no trajectory can leave $\mathcal{S}_{\varepsilon^-}$. In fact, on the side given by $v - (-\varepsilon + 1/c)u = 0$, the component of the velocity in the direction of the outward normal vector is

$$(\dot{u}, \dot{v}) \cdot \left(-\left(\frac{1}{c} - \varepsilon\right), 1\right)$$

$$= \dot{v} - \dot{u}\left(\frac{1}{c} - \varepsilon\right)$$

$$= \rho v(1 - u - v) - au - \left(\frac{1}{c} - \varepsilon\right)u(1 - u - v) + \left(\frac{1}{c} - \varepsilon\right)acu$$

$$= u\left[\left(\frac{1}{c} - \varepsilon\right)(\rho - 1)(1 - u - v) - \varepsilon ac\right]$$

$$\leqslant u\left[2\left(\frac{1}{c} - \varepsilon\right)(\rho + 1) - \varepsilon ac\right],$$

which is negative if $a \geqslant a_2$, with

$$a_2 := 2\left(\frac{1}{c} - \varepsilon\right)(\rho + 1)\frac{1}{\varepsilon c}.$$

Hence, if $(u(0), v(0)) \in \mathcal{S}_{\varepsilon^-}$, then either $T_s(u(0), v(0)) < \infty$ or $(u(t), v(t)) \in \mathcal{S}_{\varepsilon^-}$ for all $t \geqslant 0$, where the notation in (3.1) has been used. We also notice that, for $a > 1/c$, the points $(0, 1)$ and $(0, 0)$ are the only equilibria of the system, and there are no cycles. We have that $(0, 1) \notin \overline{\mathcal{S}_{\varepsilon^-}}$ and $(0, 0) \in \overline{\mathcal{S}_{\varepsilon^-}}$; thus if

$$(u(t), v(t)) \in \mathcal{S}_{\varepsilon^-} \text{ for all } t \geqslant 0, \tag{6.61}$$

then

$$(u(t), v(t)) \to (0, 0). \tag{6.62}$$

On the other hand, by Proposition 5.1, we have that at $u = 0$ the stable manifold is tangent to the line

$$v = \frac{a}{\rho - 1 + ac}u = \frac{1}{\frac{\rho-1}{a} + c}u,$$

and, if we take a large enough, this line lies above the line $v = u(1/c - \varepsilon)$. This says that, for sufficiently large t, the trajectory must lie outside $\mathcal{S}_{\varepsilon^-}$, and this is in contradiction with (6.61).

As a result of these considerations, we conclude that if

$$(u(0), v(0)) \in \mathcal{S}_{\varepsilon^-},$$

then

$$T_s(u(0), v(0)) < \infty,$$

which implies (6.60).

As a consequence of (6.60), we obtain that for every $\varepsilon \in (0, 1/c)$ there exists $a_\varepsilon > 0$ such that

$$\mathcal{S}_{\varepsilon^-} \subseteq \bigcap_{a \geq a_\varepsilon} \mathcal{E}(a).$$

In particular, for all $\varepsilon \in (0, 1/c)$, it holds that

$$\mathcal{S}_{\varepsilon^-} \subseteq \bigcap_{a' > 0} \bigcup_{a > a'} \mathcal{E}(a) = \mathcal{E}_\infty,$$

which proves (6.59), as desired.

Then, the first inclusion in (6.31) plainly follows from (6.59). $\square$

Strategies of the First Population

7

The main theorems on the winning strategy have been stated in Sect. 3.3. In particular, Theorem 3.3 gives the characterization of the set $\mathcal{V}_A$ of points that have a winning strategy in (3.7), and Theorem 3.4 establishes the nonequivalence of constant and nonconstant strategies when $\rho \neq 1$ (and their equivalence when $\rho = 1$). Nonetheless, in Theorem 3.5, we state that Heaviside functions are enough to construct a winning strategy for every point in $\mathcal{V}_A$.

In the following subsections we will give the proofs of these results.

7.1 Winning Nonconstant Strategies

We want to put in light the construction of nonconstant winning strategies for the points for which constant strategies fail.

For this, we recall the notation introduced in (3.8), (3.11) and (3.13), and we have the following statement:

Proposition 7.1 *Let $M > 1$. Then we have:*

1. *For $\rho < 1$, let (u_0, v_0) be a point in the set*

$$\mathcal{P} := \left\{ (u, v) \in [u_s^0, 1] \times [0, 1] \text{ s.t.} \right.$$
$$\left. \gamma_0(u) \leqslant v < \frac{u}{c} + \frac{1-\rho}{1+\rho c} \right\}. \tag{7.1}$$

Then there exist $a^* > M$, $a_* < \frac{1}{M}$, and $T \geqslant 0$, depending on (u_0, v_0), c, and ρ, such that for the Heaviside strategy defined by

$$a(t) = \begin{cases} a^*, & \text{if } t < T, \\ a_*, & \text{if } t \geqslant T, \end{cases} \qquad (7.2)$$

we have $(u_0, v_0) \in \mathcal{V}_\mathcal{A}$.

2. For $\rho > 1$, let (u_0, v_0) be a point of the set

$$\mathcal{Q} := \left\{ (u, v) \in [u_\infty, 1] \times [0, 1] \ \text{s.t.} \ \frac{u}{c} \leqslant v < \zeta(u) \right\}. \qquad (7.3)$$

Then there exist $a^* > M$, $a_* < \frac{1}{M}$, and $T \geqslant 0$, depending on (u_0, v_0), c, and ρ, such that for the Heaviside strategy defined by

$$a(t) = \begin{cases} 0, & \text{if } t < T, \\ a^*, & \text{if } t \geqslant T, \end{cases}$$

we have $(u_0, v_0) \in \mathcal{V}_\mathcal{A}$.

Proof We start by proving the first claim in Proposition 7.1. To this aim, we take $(\bar{u}, \bar{v}) \in \mathcal{P}$, and we observe that

$$\bar{v} - \frac{\bar{u}}{c} < \frac{1-\rho}{1+\rho c} = v_s^0 - \frac{u_s^0}{c}.$$

Therefore, there exists $\xi > 0$ such that

$$\xi < \frac{v_s^0 - \bar{v} - \frac{1}{c}(u_s^0 - \bar{u})}{\bar{u} - u_s^0}.$$

Hence, setting

$$v_S := \left(\frac{1}{c} - \xi \right) (u_s^0 - \bar{u}) + \bar{v}, \qquad (7.4)$$

we see that

$$v_S < v_s^0. \qquad (7.5)$$

Now, we want to show that there exists $a^* > 0$ such that, for any $a > a^*$ and $u > u_s^0$, we have that

7.1 Winning Nonconstant Strategies

$$\frac{\dot{v}}{\dot{u}} > \frac{1}{c} - \xi. \tag{7.6}$$

To prove this, we first notice that

$$\text{if } a > \frac{2}{c}, \text{ then } \dot{u} \leqslant -u < 0 \text{ for all } u, v \in [0, 1] \times [0, 1]. \tag{7.7}$$

Moreover, we set

$$a_1 := \frac{1+\rho c}{4c},$$

and we claim that

$$\text{if } a > a_1 \text{ and } u > u_s^0, \text{ then } \dot{v} < 0. \tag{7.8}$$

Indeed, we recall that the function σ defined in (5.5) represents the points in $[0, 1] \times [0, 1]$ where $\dot{v} = 0$ and separates the points where $\dot{v} > 0$, which lie on the left of the curve described by σ, from the points where $\dot{v} < 0$, which lie on the right of the curve described by σ.

Therefore, in order to show (7.8), it is sufficient to prove that the curve described by σ is contained in $\{u \leqslant u_s^0\}$ whenever $a > a_1$. For this, one computes that if $u = \sigma(v)$ and $a > a_1$, then

$$\begin{aligned}
u - u_s^0 &= \sigma(v) - \frac{\rho c}{1+\rho c} \\
&= 1 - \frac{\rho v^2 + a}{\rho v + a} - \frac{\rho c}{1+\rho c} \\
&= \frac{\rho v - \rho v^2}{\rho v + a} - \frac{\rho c}{1+\rho c} \\
&= \frac{\rho v(1-v)}{\rho v + a} - \frac{\rho c}{1+\rho c} \\
&\leqslant \frac{\rho}{4(\rho v + a)} - \frac{\rho c}{1+\rho c} \\
&\leqslant \frac{\rho}{4a} - \frac{\rho c}{1+\rho c} \\
&\leqslant \frac{\rho}{4a_1} - \frac{\rho c}{1+\rho c} \\
&\leqslant 0.
\end{aligned}$$

This completes the proof of (7.8).

Now we define

$$a_2 := \left(\rho + \frac{1}{c} + \xi\right) \frac{2}{u_s^0 c \xi},$$

and we claim that

$$\text{if } a > a_2 \text{ and } u > u_s^0, \text{ then } \dot{v} < \left(\frac{1}{c} - \xi\right) \dot{u}. \tag{7.9}$$

Indeed, under the assumptions of (7.9), we deduce that

$$\dot{v} - \left(\frac{1}{c} - \xi\right) \dot{u}$$

$$= \rho v(1 - u - v) - au - \left(\frac{1}{c} - \xi\right)\left(u(1 - u - v) - acu\right)$$

$$= (1 - u - v)\left(\rho v - \left(\frac{1}{c} - \xi\right)u\right) - ac\xi u$$

$$\leqslant 2\left(\rho v + \frac{u}{c} + \xi u\right) - ac\xi u$$

$$< 2\left(\rho + \frac{1}{c} + \xi\right) - a_2 c \xi u_s^0$$

$$= 0,$$

and this establishes the claim in (7.9).

Then, choosing

$$a^* := \max\left\{\frac{2}{c}, a_1, a_2, M\right\},$$

we can exploit (7.7), (7.8), and (7.9) to deduce (7.6), as desired.

Now we claim that, for any $a > a^*$, there exists $T \geqslant 0$ such that the trajectory $(u(t), v(t))$ starting from $(\bar{u}, \bar{v})$ satisfies

$$u(T) = u_s^0 \text{ and } v(T) < v_S. \tag{7.10}$$

Indeed, we define $T \geqslant 0$ to be the first time for which $u(T) = u_s^0$. This is a fair definition, since $u(0) = \bar{u} \geqslant u_s^0$ and $\dot{u}$ is negative and bounded away from zero till $u \geqslant u_s^0$, thanks to (7.7). Then, we see that

$$v(T) = \bar{v} + \int_0^T \dot{v}(t) \, dt$$

7.1 Winning Nonconstant Strategies

$$< \bar{v} + \int_0^T \left(\frac{1}{c} - \xi\right) \dot{u}(t)\, dt$$

$$= \bar{v} + \left(\frac{1}{c} - \xi\right)(u(T) - u(0))$$

$$= \bar{v} + \left(\frac{1}{c} - \xi\right)(u_s^0 - \bar{u})$$

$$= v_S,$$

thanks to (7.4) and (7.6), and this establishes (7.10).

Now we observe that

$$v(T) < v_S < v_s^0 = \gamma_0(u_s^0) = \gamma_0(u(T))$$

due to (7.5) and (7.10).

As a result, recalling Lemma 6.5, we can choose $a_* < 1/M$ such that

$$v(T) < \gamma_{a_*}(u(T)).$$

Accordingly, by Proposition 5.9, we obtain that $(u(T), v(T)) \in \mathcal{E}(a_*)$. Hence, applying the strategy in (7.2), we accomplish the desired result and complete the proof of the first claim in Proposition 7.1.

Now we focus on the proof of the second claim in Proposition 7.1. For this, let

$$(u_0, v_0) \in \mathcal{Q}, \tag{7.11}$$

and consider the trajectory $(u_0(t), v_0(t))$ starting from (u_0, v_0) for the strategy $a = 0$. In light of formula (6.24) of Proposition 6.3, we have that

$$\begin{aligned}&\text{the trajectory } (u_0(t), v_0(t)) \text{ converges} \\ &\text{to a point of the form } (u_F, 1 - u_F) \text{ as } t \to +\infty.\end{aligned} \tag{7.12}$$

We define

$$v_F := 1 - u_F, \quad v_\infty := 1 - u_\infty = \frac{1}{c+1}, \tag{7.13}$$

where the last equality can be checked starting from the value of u_∞ given in (3.13). Using the definition of ζ in (3.13), we have

$$\zeta(u_\infty) = \frac{1}{c(u_\infty)^{\rho-1}} u_\infty^\rho = \frac{c}{c(c+1)} = v_\infty,$$

and, recalling (7.13) and formula (6.24) of Proposition 6.3, we get that the graph of ζ is the union of trajectories for $a = 0$ that converges to $(u_\infty, 1 - u_\infty)$ as $t \to +\infty$.

Also, by (7.11), we have that $v_0 < \zeta(u_0)$. Thus, since by Cauchy's uniqueness result for ODEs, two orbits never intersect, we have that

$$\text{the orbit } (u_0(t), v_0(t)) \text{ must lie below the graph of } \zeta. \tag{7.14}$$

Since both (u_F, v_F) and (u_∞, v_∞) belong to the line given by $v = 1 - u$, from (7.14) we get that

$$u_\infty < u_F \tag{7.15}$$

and

$$v_\infty > v_F. \tag{7.16}$$

Thanks to (7.15) and (7.16) and recalling the values of u_∞ from (3.13) and of v_∞ from (7.13), we get that

$$v_F < v_\infty = \frac{u_\infty}{c} < \frac{u_F}{c}. \tag{7.17}$$

As a consequence, since the inequality in (7.17) is strict, there exists $T' > 0$ such that

$$v_0(T') < \frac{u_0(T')}{c}. \tag{7.18}$$

Moreover, since $\dot{u} < 0$ for $v > 1 - u$ and $a = 0$, we get that $u_0(t)$ is decreasing in t, and therefore $u_F < u_0(T') < u_0$.

By the strict inequality in (7.18), and claim (ii) in Proposition 6.4, we have that $(u_0(T'), v_0(T')) \in \mathcal{E}_\infty$, where $\mathcal{E}_\infty$ is defined in (6.29). In particular, we have that

$$(u_0(T'), v_0(T')) \in \bigcup_{a > a'} \mathcal{E}(a),$$

for every $a' > 0$.

Consequently, there exists $a^* > M$ such that

$$(u_0(T'), v_0(T')) \in \mathcal{E}(a^*).$$

Therefore, applying the strategy

$$a(t) = \begin{cases} 0, & t < T', \\ a^*, & t \geq T, \end{cases}$$

we reach the claim $(u_0, v_0) \in \mathcal{V}_\mathcal{A}$. $\square$

7.2 Winning Strategies

To avoid repeating passages in the proofs of Theorems 3.3 and 3.4, we first state and prove the following lemma:

Lemma 7.2 *If $\rho = 1$, then for all $a > 0$:*

1. The curve

$$v(\tau) = \frac{e^{c\tau}\left(c\frac{v(0)}{u(0)} - 1\right) + 1}{c} u(\tau) \quad \text{for } \tau > 0$$

is a parametrization of the trajectory starting in $(u(0), v(0))$.
2. We have $\mathcal{E}(a) = \mathcal{S}_c$, where $\mathcal{S}_c$ is defined in (6.32).

Proof Let $(u(t), v(t))$ be a trajectory starting at a point in $[0, 1] \times [0, 1]$. For any $a > 0$, we consider the function

$$\mu(t) := \frac{v(t/a)}{u(t/a)}.$$

Notice that

$$\begin{aligned}(u(0), v(0)) \in \mathcal{E}(a) \text{ if and only if} \\ \text{there exists } T > 0 \text{ such that } \mu(T) = 0.\end{aligned} \quad (7.19)$$

In addition, we observe that

$$\begin{aligned}\dot{\mu}(t) &= \frac{\dot{v}(t/a)\, u(t/a) - v(t/a)\, \dot{u}(t/a)}{au^2(t/a)} \\ &= \frac{-u^2(t/a) + cu(t/a)\, v(t/a)}{u^2(t/a)} \\ &= c\mu(t) - 1.\end{aligned}$$

We deduce that

$$\mu(t) = \frac{e^{ct}(c\mu(0) - 1) + 1}{c}. \quad (7.20)$$

This proves the first part of the Lemma.

From (7.20) and (7.19), we deduce that

$$(u(0), v(0)) \in \mathcal{E}(a) \text{ if and only if } c\mu(0) - 1 < 0.$$

This leads to

$$(u(0), v(0)) \in \mathcal{E}(a) \text{ if and only if } \frac{v(0)}{u(0)} < \frac{1}{c},$$

which, recalling the definition of $\mathcal{S}_c$ in (6.32), ends the proof. □

Now we provide the proof of Theorem 3.3, exploiting the result obtained in Sect. 7.1.

Proof of Theorem 3.3

(i) Let $\rho = 1$. We claim that

$$\mathcal{V}_A = \mathcal{S}_c, \quad (7.21)$$

where $\mathcal{S}_c$ was defined in (6.32) (incidentally, $\mathcal{S}_c$ is precisely the right-hand side of Eq. (3.9)).

From Lemma 7.2 we have that for $\rho = 1$ and $a > 0$ it holds $\mathcal{S}_c = \mathcal{E}(a) \subseteq \mathcal{V}_A$. Thus, to show (7.21) we just need to check that

$$\mathcal{V}_A \subseteq \mathcal{S}_c, \quad (7.22)$$

which is equivalent to

$$\mathcal{S}_c^C \subseteq \mathcal{V}_A^C, \quad (7.23)$$

where the superscript C denotes the complement of the set in the topology of $[0, 1] \times [0, 1]$.

In order to apply Lemma (4.7), we analyze the behavior of the trajectories at $\partial \mathcal{S}_c^C$. It holds that

$$\partial \mathcal{S}_c^C \cap ([0, 1] \times [0, 1]) = \left\{ (u, v) \in (0, 1) \times (0, 1) \text{ s.t. } v = \frac{u}{c} \right\}.$$

7.2 Winning Strategies

Then, let us analyze the curve $v = \frac{u}{c}$ for $u \in [0, u_M]$ where $u_M = \min\{1, c\}$. By Lemma 7.2, for all $a > 0$ and by Proposition 6.3 if $a = 0$, the considered curve is a parametrization of a trajectory. Thus, for $\check{u} \in [0, u_M]$, for all points in the form $(\check{u}, \check{v}) = (\check{u}, \check{u}/c)$, it holds that the trajectory $(u(t), v(t))$ starting at $(\check{u}, \check{v})$ at $t = 0$ satisfies

$$(\dot{u}(0), \dot{v}(0)) \cdot \nu = 0 \quad \text{for } \check{u} \in [0, u_M], \check{v} = \check{u}/c, \tag{7.24}$$

where ν is the outward unit normal vector to ∂S_c^C at $(\check{u}, \check{v})$.

Hence, by choosing $K_1 = [0, u_M]$, thanks to (7.24), we can apply Lemma 4.7 and conclude that

$$\text{no trajectory can exit } S_c^C.$$

Next, we observe that

$$S_c^C \cap ((0, 1] \times \{0\}) = \varnothing,$$

so no trajectory starting in S_c^C can reach the set $(0, 1] \times \{0\}$.
Therefore,

$$S_c^C \cap \mathcal{V}_\mathcal{A} = \varnothing,$$

and this implies that (7.23) is true. As a result, the proof of (7.22) is established, and the proof is completed for $\rho = 1$.

(ii) Let $\rho < 1$. Let $\mathcal{Y}$ be the set in the right-hand side of (3.10), and

$$\mathcal{F}_0 := \big\{ (u, v) \in [0, 1] \times [0, 1] \text{ s.t. } v < \gamma_0(u) \text{ if } u \in [0, 1] \big\}. \tag{7.25}$$

Notice that

$$\mathcal{Y} = \mathcal{F}_0 \cup \mathcal{P}, \tag{7.26}$$

being $\mathcal{P}$ the set defined in (7.1).
Moreover,

$$\mathcal{P} \subseteq \mathcal{V}_\mathcal{A}, \tag{7.27}$$

thanks to Proposition 7.1.
We also claim that

$$\mathcal{F}_0 \subseteq \mathcal{V}_\mathcal{K}, \tag{7.28}$$

where $\mathcal{K}$ is the set of constant functions. Indeed, if $(u, v) \in \mathcal{F}_0$, we have that $v < \gamma_0(u)$ and consequently $v < \gamma_a(u)$, as long as a is small enough, due to Lemma 6.5. From this and Proposition 5.9, we deduce that (u, v) belongs to $\mathcal{E}(a)$, as long as a is small enough, and this proves (7.28).

From (7.28) and the fact that $\mathcal{K} \subseteq \mathcal{A}$, we obtain that

$$\mathcal{F}_0 \subseteq \mathcal{V}_\mathcal{A}. \tag{7.29}$$

Then, as a consequence of (7.26), (7.27), and (7.29), we get that $\mathcal{Y} \subseteq \mathcal{V}_\mathcal{A}$. Hence, we are left with proving that

$$\mathcal{V}_\mathcal{A} \subseteq \mathcal{Y}. \tag{7.30}$$

For this, we show that

$$\begin{array}{c} \text{no trajectory enters } \mathcal{Y} \\ \text{through } \partial \mathcal{Y} \cap ([0, 1] \times [0, 1]). \end{array} \tag{7.31}$$

To prove this, in order to be able to apply Lemma 4.6, we calculate the outward normal derivative on the part of $\partial \mathcal{Y}$ lying on the graph of $v = \gamma_0(u)$ for $u \in [0, u_s]$, that is, up to a positive constant of normalization,

$$\dot{v} - \frac{\rho v_s^0 u^{\rho-1} \dot{u}}{(u_s^0)^\rho} = \rho v(1 - u - v) - au - \frac{\rho v_s^0 u^\rho (1 - u - v - ac)}{(u_s^0)^\rho}.$$

By substituting for $v = \gamma_0(u) = \frac{v_s^0 u^\rho}{(u_s^0)^\rho}$, we get

$$\dot{v} - \frac{\rho v_s^0 u^{\rho-1} \dot{u}}{(u_s^0)^{\rho-1}}$$

$$= \frac{\rho v_s^0 u^\rho}{(u_s^0)^\rho}(1 - u - v) - au - \frac{\rho v_s^0 u^\rho (1 - u - v - ac)}{(u_s^0)^\rho}$$

$$= -au + \frac{ac\rho v_s^0 u^\rho}{(u_s^0)^\rho}$$

$$= au^\rho \left(-u^{1-\rho} + \frac{c\rho v_s^0}{(u_s^0)^\rho} \right)$$

$$= au^\rho \left(-u^{1-\rho} + \frac{1}{(u_s^0)^{\rho-1}} \right).$$

7.2 Winning Strategies

As a result, since $\rho < 1$, we have

$$\dot{v} - \frac{\rho v_s^0 u^{\rho-1} \dot{u}}{(u_s^0)^\rho} > 0 \quad \text{for } u \in (0, u_s) \tag{7.32}$$

and

$$\dot{v} - \frac{\rho v_s^0 u^{\rho-1} \dot{u}}{(u_s^0)^\rho} = 0 \quad \text{for } u \in \{0, u_s\}. \tag{7.33}$$

Notice that $\partial \mathcal{Y}$ coincides on the line $v = \frac{u}{c} + \frac{1-\rho}{1+\rho c}$ for

$$u \in \left[u_s^0, \min\left\{1, \frac{\rho c(c+1)}{1+\rho c}\right\} \right].$$

For the sake of simplicity, we suppose that

$$\frac{\rho c(c+1)}{1+\rho c} \geqslant 1.$$

Then, on the part of $\partial \mathcal{Y}$ contained on the line $v = \frac{u}{c} + \frac{1-\rho}{1+\rho c}$, the outward normal derivative is, up to the constant $1/\sqrt{1+(1/c^2)}$,

$$\dot{v} - \frac{\dot{u}}{c}$$

$$= \rho v(1-u-v) - au - \frac{u(1-ac-u-v)}{c}$$

$$= \left(\rho v - \frac{u}{c}\right)(1-u-v) \tag{7.34}$$

$$= \left(\frac{\rho u}{c} + \frac{\rho(1-\rho)}{1+\rho c} - \frac{u}{c}\right)\left(1 - u - \frac{u}{c} - \frac{1-\rho}{1+\rho c}\right)$$

$$= \left(\frac{(\rho-1)u}{c} + \frac{\rho(1-\rho)}{1+\rho c}\right)\left(-\frac{u(c+1)}{c} + \frac{\rho(1+c)}{1+\rho c}\right).$$

We also observe that, when $u > u_s^0 = \frac{\rho c}{1+\rho c}$, the condition $\rho < 1$ gives that

$$\frac{(\rho-1)u}{c} + \frac{\rho(1-\rho)}{1+\rho c} < \frac{\rho(\rho-1)}{1+\rho c} + \frac{\rho(1-\rho)}{1+\rho c} = 0$$

and

$$-\frac{u(c+1)}{c} + \frac{\rho(1+c)}{1+\rho c} < -\frac{\rho(c+1)}{1+\rho c} + \frac{\rho(1+c)}{1+\rho c} = 0.$$

Therefore, when $u > u_s^0$, we deduce from (7.34) that

$$\dot{v} - \frac{\dot{u}}{c} > 0,$$

and

$$\dot{v} - \frac{\dot{u}}{c} = 0 \quad \text{for } u = u_s^0.$$

Combining this, (7.32) and (7.32), we can say that for $K_1 = \{0\}$, $K_2 = \{u_s^0\}$, $I_1 = (0, u_s^0)$, $I_2 = (u_s^0, 1]$, we can apply Lemma 4.6 obtaining (7.31), as desired.

Since for any value of a, no trajectory starting in $([0, 1] \times [0, 1]) \setminus \mathcal{Y}$ can enter in $\mathcal{Y}$ passing through $\partial \mathcal{Y} \cap ([0, 1] \times [0, 1])$, in particular no trajectory starting in $([0, 1] \times [0, 1]) \setminus \mathcal{Y}$ can hit $\{v = 0\}$, which ends the proof of (7.30).

(iii) Let $\rho > 1$. For the sake of simplicity, we suppose that

$$\frac{c}{(c+1)^\rho} \geq 1.$$

Let $\mathcal{X}$ be the right-hand side of (3.12). We observe that

$$\mathcal{X} = \mathcal{S}_c \cup \mathcal{Q}, \tag{7.35}$$

where $\mathcal{S}_c$ was defined in (6.32) and $\mathcal{Q}$ in (7.3).

Thanks to Proposition 6.4, one has that

$$\mathcal{S}_c \subseteq \bigcup_{a > a'} \mathcal{E}(a),$$

for every $a' > 0$, and therefore $\mathcal{S}_c \subseteq \mathcal{V}_\mathcal{A}$.

Moreover, by the second claim in Proposition 7.1, one also has that $\mathcal{Q} \subseteq \mathcal{V}_\mathcal{A}$. Hence,

$$\mathcal{X} \subseteq \mathcal{V}_\mathcal{A}. \tag{7.36}$$

Accordingly, to prove equality in (7.36) and thus complete the proof of (3.12), we need to show that

$$\mathcal{V}_\mathcal{A} \subseteq \mathcal{X}. \tag{7.37}$$

First, we prove that

$$(0, 1] \times \{0\} \subseteq \mathcal{X}. \tag{7.38}$$

7.2 Winning Strategies

Indeed, for $u > 0$ we have $v = \frac{u}{c} > 0$; therefore $(u, 0) \in \mathcal{X}$ for $u \in (0, u_\infty]$. Then, $\zeta(u)$ is increasing in u since it is a positive power function, therefore $v = \zeta(u) > 0$ for $u \in (u_\infty, 1]$, and hence $(u, 0) \in \mathcal{X}$ for $u \in (u_\infty, 1]$. These observations prove (7.38).

Now, to show that

$$\text{no trajectory enters } \mathcal{X}, \tag{7.39}$$

we want to apply Lemma 4.6 with $K_1 = \{0\}$, $I_1 = (0, u_\infty)$, $K_2 = \{u_\infty\}$, $I_2 = (u_\infty, 1]$.

First, we notice that $(0, 0) \in \partial \mathcal{X}$ is an equilibrium, and thus the component of the velocity is 0 in every direction. This proves (4.3) for $K_1 = \{0\}$.

We now prove that the component of the velocity field in the outward normal direction with respect to $\mathcal{X}$ is positive on

$$\left\{ (u, v) \in (0, u_\infty) \times (0, 1) : v = \frac{u}{c} \right\}.$$

In fact, it is

$$\begin{aligned}\dot{v} - \frac{1}{c}\dot{u} &= \rho v(1 - u - v) - au - \frac{u}{c}(1 - ac - u - v) \\ &= \left(\rho v - \frac{u}{c}\right)(1 - u - v).\end{aligned} \tag{7.40}$$

The first term is positive because for $\rho > 1$ we have

$$\rho v > v = \frac{u}{c}.$$

Moreover, for $u < u_\infty$, we have that

$$1 - u - v > 1 - u_\infty - \frac{u_\infty}{c} = 0,$$

thanks to (3.13). Thus, the left-hand side of (7.40) is positive for $u \in I_1 = (0, u_\infty)$.

On the part of $\partial \mathcal{X}$ lying in the graph of $v = \zeta(u)$, that is, for $u \in I_2$, the component of the velocity field in the outward normal direction is given by

$$\begin{aligned}\dot{v} &- \frac{\rho u^{\rho-1} \dot{u}}{\rho c (u_\infty)^{\rho-1}} \\ &= \rho v(1 - u - v) - au - \frac{\rho u^\rho}{\rho c (u_\infty)^{\rho-1}}(1 - u - v - ac).\end{aligned} \tag{7.41}$$

Now we substitute for

$$v = \zeta(u) = \frac{u^\rho}{\rho c(u_\infty)^{\rho-1}}$$

in (7.41), and we get

$$\dot{v} - \frac{u^{\rho-1}\dot{u}}{c(u_\infty)^{\rho-1}} = au\left(-1 + \frac{u^{\rho-1}}{(u_\infty)^{\rho-1}}\right), \qquad (7.42)$$

which leads to

$$\dot{v} - \frac{\rho u^{\rho-1}\dot{u}}{\rho c(u_\infty)^{\rho-1}} > 0 \quad \text{if } u > u_\infty,$$

as desired. This proves the hypothesis (4.2) for $u \in I_2$.

Moreover, again by the expression in (7.40) and to (3.13), for $u = u_\infty$, the scalar product of the direction of the trajectory with the normal vector to $\{v = \frac{u}{c}\}$ is zero. Also, by (7.42), for $u = u_\infty$, the scalar product of the direction of the trajectory with the normal vector to $\{v = \zeta(u)\}$ is zero. Thus, the hypothesis (4.3) of Lemma 4.6 is satisfied in $K_2 = \{u_\infty\}$.

As a consequence of these considerations, by Lemma 4.6 we find that no trajectory starting in $\mathcal{X}^C$ can enter in $\mathcal{X}$ and therefore hit $\{v = 0\}$, by (7.38).

Hence, we conclude that (7.37) holds true, which, together with (7.36), establishes (3.12). □

7.3 The Role of the Constant Strategies

In order to prove Theorem 3.4, we will establish a geometrical lemma in order to understand the reciprocal position of the function γ, as given by Propositions 5.1 and 5.7, and the straight line where the saddle equilibria lie. To emphasize the dependence of γ on the parameter a, we will often use the notation $\gamma = \gamma_a$. Moreover, we recall the notation of the saddle points (u_s, v_s) defined in (3.5) and of the points $(u_\mathcal{M}, v_\mathcal{M})$ given by Propositions 5.1 and 5.7, with the convention that

$$(u_s, v_s) = (0, 0) \text{ if } ac \geqslant 1, \qquad (7.43)$$

and we state the following result:

Lemma 7.3 *If $\rho < 1$, then*

$$\frac{u}{\rho c} \leqslant \gamma_a(u) \quad \text{for } u \in [0, u_s] \qquad (7.44)$$

7.3 The Role of the Constant Strategies

and

$$\gamma_a(u) \leqslant \frac{u}{\rho c} \quad \text{for } u \in [u_s, u_\mathcal{M}]. \tag{7.45}$$

If instead $\rho > 1$, then

$$\gamma_a(u) \leqslant \frac{u}{\rho c} \quad \text{for } u \in [0, u_s] \tag{7.46}$$

and

$$\frac{u}{\rho c} \leqslant \gamma_a(u) \quad \text{for } u \in [u_s, u_\mathcal{M}]. \tag{7.47}$$

Moreover, equality holds in (7.44) and (7.46) if and only if either $u = u_s$ or $u = 0$. Also, strict inequality holds in (7.45) and (7.47) for $u \in (u_s, u_\mathcal{M})$.

Proof We focus here on the proof of (7.45), since the other inequalities are proven in a similar way. Moreover, we deal with the case $ac < 1$, being the case $ac \geqslant 1$ analogous with obvious modifications.

We suppose by contradiction that (7.45) does not hold true. Namely, we assume that there exists $\tilde{u} \in (u_s, u_\mathcal{M}]$ such that

$$\gamma_a(\tilde{u}) > \frac{\tilde{u}}{\rho c}.$$

Since γ_a is continuous thanks to Propositions 5.1, we have that

$$\gamma_a(u) > \frac{u}{\rho c} \quad \text{in a neighborhood of } \tilde{u}.$$

Hence, we consider the largest open interval $(u_1, u_2) \subset (u_s, u_\mathcal{M}]$ containing $\tilde{u}$ and such that

$$\gamma_a(u) > \frac{u}{\rho c} \quad \text{for all } u \in (u_1, u_2). \tag{7.48}$$

Moreover, in light of (3.5), we see that

$$\gamma_a(u_s) = v_s = \frac{1 - ac}{1 + \rho c} = \frac{u_s}{\rho c}. \tag{7.49}$$

Hence, by the continuity of γ_a, we have that $\gamma_a(u_1) = \frac{u_1}{\rho c}$ and

$$\text{either } \gamma_a(u_2) = \frac{u_2}{\rho c} \text{ or } u_2 = u_\mathcal{M}. \tag{7.50}$$

Now, we consider the set

$$\mathcal{T} := \left\{ (u, v) \in [u_1, u_2] \times [0, 1] \text{ s.t. } \frac{u}{\rho c} < v < \gamma_a(u) \right\},$$

that is, nonempty, thanks to (7.48). We claim that

$$\text{for all } (u(0), v(0)) \in \mathcal{T}, \quad \text{the } \omega\text{-limit of its trajectory is } (u_s, v_s). \tag{7.51}$$

To prove this, we analyze the normal derivative on

$$\partial \mathcal{T} = \mathcal{T}_1 \cup \mathcal{T}_2 \cup \mathcal{T}_3,$$

where $\mathcal{T}_1 := \{(u, \gamma_a(u)) \text{ with } u \in (u_1, u_2)\}$,

$$\mathcal{T}_2 := \left\{ \left(u, \frac{u}{\rho c}\right) \text{ with } u \in (u_1, u_2) \right\}$$

and $\mathcal{T}_3 := \left\{ (u_2, v) \text{ with } v \in \left(\frac{u_2}{\rho c}, \min\{\gamma_a(u_2), 1\} \right) \right\}$,

with the convention that $\partial \mathcal{T}$ does contain $\mathcal{T}_3$ only if the second possibility in (7.50) occurs.

We notice that the set $\mathcal{T}_1$ is an orbit for the system, and thus the component of the velocity in the normal direction is null. On $\mathcal{T}_2$, we have that the sign of the component of the velocity in the inward normal direction is given by

$$(\dot{u}, \dot{v}) \cdot \left(-\frac{1}{\rho c}, 1\right)$$

$$= \dot{v} - \frac{1}{\rho c} \dot{u}$$

$$= \rho v (1 - u - v) - au - \frac{u}{\rho c}(1 - u - v) + \frac{au}{\rho} \tag{7.52}$$

$$= \frac{u}{c}\left(1 - u - \frac{u}{\rho c}\right)\left(1 - \frac{1}{\rho}\right) - au\left(1 - \frac{1}{\rho}\right)$$

$$= \frac{u}{c}\left(1 - \frac{1}{\rho}\right)\left(1 - u - \frac{u}{\rho c} - ac\right).$$

Notice that for $u \geqslant u_s$ we have that

$$1 - u - v - ac \leqslant 0, \tag{7.53}$$

and thus the sign of last term in (7.52) depends only on the quantity $1 - \frac{1}{\rho}$. Consequently, since $\rho < 1$, the sign of the component of the velocity in the inward normal direction is positive.

7.3 The Role of the Constant Strategies

Furthermore, in the case in which the second possibility in (7.50) occurs, we also check the sign of the component of the velocity in the inward normal direction along $\mathcal{T}_3$. In this case, if $\gamma_a(u_2) < 1$, then $u_2 = 1$, and therefore we find that

$$(\dot{u}, \dot{v}) \cdot (-1, 0) = -\dot{u} = -u(1 - u - v) + acu = v + ac,$$

which is positive. If instead $\gamma_a(u_2) = 1$

$$(\dot{u}, \dot{v}) \cdot (-1, 0) = -\dot{u} = -u(1 - u - v) + acu = -u(1 - ac - u - v),$$

which is positive, thanks to (7.53).

We also point out that there are no cycles in $\mathcal{T}$, since $\dot{u}$ has a sign. These considerations and the Poincaré–Bendixson Theorem (see, e.g., [90]) give that the ω-limit set of $(u(0), v(0))$ can be either an equilibrium or a union of (finitely many) equilibria and non-closed orbits connecting these equilibria. Since $(0, 0)$ and $(0, 1)$ do not belong to the closure of $\mathcal{T}$, in this case the only possibility is that the ω-limit is the equilibrium (u_s, v_s). Consequently, we have that $u_1 = u_s$ and that (7.51) is satisfied.

Accordingly, in light of (7.51), we have that the set $\mathcal{T}$ is contained in the stable manifold of (u_s, v_s), which is in contradiction with the definition of $\mathcal{T}$. Hence, (7.45) is established, as desired.

Now we show that strict inequality holds true in (7.45) if $u \in (u_s, u_\mathcal{M})$. To this end, we suppose by contradiction that there exists $\bar{u} \in (u_s, u_\mathcal{M})$ such that

$$\gamma_a(\bar{u}) = \frac{\bar{u}}{\rho c}. \tag{7.54}$$

Now, since (7.45) holds true, we have that the line $v - \frac{u}{\rho c} = 0$ is tangent to the curve $v = \gamma_a(u)$ at $(\bar{u}, \gamma_a(\bar{u}))$, and therefore at this point the components of the velocity along the normal directions to the curve and to the line coincide. On the other hand, the normal derivative at a point on the line has a sign, as computed in (7.52), while the normal derivative to $v = \gamma_a(u)$ is 0 because the curve is an orbit.

This, together with (7.49), proves that equality in (7.45) holds true if $u = u_s$, but strict inequality holds true for all $u \in (u_s, u_\mathcal{M})$, and thus the proof of Lemma 7.3 is complete. □

For each $a > 0$, we define $(u_d^a, v_d^a) \in [0, 1] \times [0, 1]$ as the unique intersection of the graph of γ_a with the line $\{v = 1 - u\}$, that is, the solution of the system

$$\begin{cases} v_d^a = \gamma_a(u_d^a), \\ v_d^a = 1 - u_d^a. \end{cases} \tag{7.55}$$

We recall that the above intersection is unique since the function γ_a is increasing. Also, by construction,

$$u_d^a \leqslant u_\mathcal{M}. \tag{7.56}$$

Now, recalling (3.5) and making explicit the dependence on a by writing u_s^a (with the convention in (7.43)), we give the following result:

Lemma 7.4 *We have that:*

1. *For $\rho < 1$, for all $a^* > 0$, it holds that*

$$\gamma_a(u) \leqslant \gamma_{a^*}(u) \tag{7.57}$$
$$\text{for all } a > a^* \text{ and for all } u \in [u_s^{a^*}, u_d^{a^*}].$$

2. *For $\rho > 1$, for all $a^* > 0$, it holds that*

$$\gamma_a(u) \leqslant \gamma_{a^*}(u) \tag{7.58}$$
$$\text{for all } a < a^* \text{ and for all } u \in [u_s^{a^*}, u_d^{a^*}].$$

Proof We claim that

$$u_s^{a^*} < u_d^{a^*}. \tag{7.59}$$

Indeed, when $a^*c \geqslant 1$, we have that $u_s^{a^*} = 0 < u_d^{a^*}$, and thus (7.59) holds true. If instead $a^*c < 1$, by (3.5) and (7.55), we have that

$$\gamma_{a^*}(u_s^{a^*}) + u_s^{a^*} = 1 - a^*c < 1 = \gamma_{a^*}(u_d^{a^*}) + u_d^{a^*}. \tag{7.60}$$

Also, since γ_{a^*} is increasing, we have that the map $r \mapsto \gamma_{a^*}(r) + r$ is strictly increasing. Consequently, we deduce from (7.60) that (7.59) holds true in this case as well.

Now we suppose that $\rho < 1$ and we prove (7.57). For this, we claim that, for every $a^* > 0$ and every $a > a^*$,

$$\gamma_a(u_s^{a^*}) \leqslant \gamma_{a^*}(u_s^{a^*})$$
$$\text{with strict inequality when } a^* \in \left(0, \frac{1}{c}\right). \tag{7.61}$$

7.3 The Role of the Constant Strategies

To check this, we distinguish two cases. If $a^* \in \left(0, \frac{1}{c}\right)$, then for all $a > a^*$

$$u_s^a = \max\left\{0, \rho c \frac{1-ac}{1+\rho c}\right\} < \rho c \frac{1-a^*c}{1+\rho c} = u_s^{a^*}. \tag{7.62}$$

By (7.62) and formula (7.45) in Lemma 7.3, we have that

$$\gamma_a(u_s^{a^*}) < \frac{u_s^{a^*}}{\rho c} = \gamma_{a^*}(u_s^{a^*}) \quad \text{for all } a > a^*. \tag{7.63}$$

If instead $a^* \geq \frac{1}{c}$, then $u_s^{a^*} = 0$, and for all $a > a^*$, we have $u_s^a = 0$. As a consequence,

$$\gamma_{a^*}(u_s^{a^*}) = \gamma_a(u_s^{a^*}) \quad \text{for all } a > a^*. \tag{7.64}$$

The claim in (7.61) thus follows from (7.63) and (7.64).

Furthermore, by Propositions 5.1 and 5.7,

$$\gamma'_a(0) = \frac{a}{\rho + ac - 1} < \frac{a^*}{\rho + a^*c - 1} = \gamma'_{a^*}(0) \tag{7.65}$$

for all $a > a^* \geq \frac{1}{c}$.

Moreover, for all $a \geq a^*$ and $u > u_s^{a^*}$, it holds that, when $v = \gamma_{a^*}(u)$,

$$-(acu - u(1-u-v)) = u(1-u-\gamma_{a^*}(u) - ac)$$
$$< u(1 - u_s^{a^*} - v_s^{a^*} - ac) \tag{7.66}$$
$$\leq 0.$$

Now, we establish that

$$u(\rho cv - u)(1-u-v)(a-a^*) < 0 \tag{7.67}$$

for all $a > a^*$, $u \in (u_s^{a^*}, u_d^{a^*})$, $v = \gamma_{a^*}(u)$.

Indeed, for the values of a, u, and v as in (7.67), we have that $v \leq \gamma_{a^*}(u_d^{a^*})$ and hence

$$(1-u-v) > (1 - u_d^{a^*} - \gamma_{a^*}(u_d^{a^*})) = 0. \tag{7.68}$$

Moreover, by formula (7.45) in Lemma 7.3, for $u \in (u_s^{a^*}, u_d^{a^*})$ and $v = \gamma_{a^*}(u)$, we have that

$$\rho c v - u = \rho c \gamma_{a^*}(u) - u < 0.$$

From this and (7.68), we see that (7.67) plainly follows, as desired.

As a consequence of (7.66) and (7.67), one deduces that, for all $a > a^*$, $u \in (u_s^{a^*}, u_d^{a^*})$ and $v = \gamma_{a^*}(u)$,

$$\begin{aligned}
& \frac{au - \rho v(1-u-v)}{acu - u(1-u-v)} - \frac{a^* u - \rho v(1-u-v)}{a^* cu - u(1-u-v)} \\
&= \frac{(a-a^*)c\rho uv(1-u-v) - (a-a^*)u^2(1-u-v)}{\bigl(acu - u(1-u-v)\bigr)\bigl(a^* cu - u(1-u-v)\bigr)} \\
&= \frac{(a-a^*)(1-u-v)u(c\rho v - u)}{\bigl(acu - u(1-u-v)\bigr)\bigl(a^* cu - u(1-u-v)\bigr)} \\
&\leqslant 0.
\end{aligned} \qquad (7.69)$$

Now, we define

$$\mathcal{Z}(u) := \gamma_a(u) - \gamma_{a^*}(u) \qquad (7.70)$$

and we claim that

$$\text{if } \check{u} \in (u_s^{a^*}, u_d^{a^*}) \text{ is such that } \mathcal{Z}(\check{u}) = 0, \text{ then } \mathcal{Z}'(\check{u}) < 0. \qquad (7.71)$$

Indeed, since γ_a is a trajectory for (2.1), if $(u_a(t), v_a(t))$ is a solution of (2.1), we have that $v_a(t) = \gamma_a(u_a(t))$, whence

$$\begin{aligned}
& \rho v_a(t)(1 - u_a(t) - v_a(t)) - a u_a(t) \\
&= \dot{v}_a(t) \\
&= \gamma_a'(u_a(t))\, \dot{u}_a(t) \\
&= \gamma_a'(u_a(t))\bigl(u_a(t)(1 - u_a(t) - v_a(t)) - a c u_a(t)\bigr).
\end{aligned} \qquad (7.72)$$

Then, we let $\check{v} := \gamma_a(\check{u})$, and we notice that $\check{v}$ coincides also with $\gamma_{a^*}(\check{u})$. Hence, we take trajectories of the system with parameter a and a^* starting at $(\check{u}, \check{v})$, and by (7.69) we obtain that

7.3 The Role of the Constant Strategies

$$0 > \frac{a\check{u} - \rho v(1-\check{u}-\check{v})}{ac\check{u} - \check{u}(1-\check{u}-\check{v})} - \frac{a^*\check{u} - \rho v(1-\check{u}-\check{v})}{a^*c\check{u} - \check{u}(1-\check{u}-\check{v})}$$

$$= \frac{au_a(0) - \rho v(1 - u_a(0) - v_a(0))}{acu_a(0) - u(1 - u_a(0) - v_a(0))}$$

$$\quad - \frac{a^* u_{a^*}(0) - \rho v(1 - u_{a^*}(0) - v_a(0))}{a^* c u_{a^*}(0) - u(1 - u_{a^*}(0) - v_{a^*}(0))}$$

$$= \gamma'_a(u_a(0)) - \gamma'_{a^*}(u_{a^*}(0))$$

$$= \gamma'_a(\check{u}) - \gamma'_{a^*}(\check{u}),$$

which establishes (7.71).

Now we claim that

$$\text{there exists } \underline{u} \in [u_s^{a^*}, u_d^{a^*}] \text{ such that } \mathcal{Z}(\underline{u}) < 0 \quad\quad (7.73)$$
$$\text{and } \mathcal{Z}(u) \leqslant 0 \text{ for every } u \in [u_s^{a^*}, \underline{u}].$$

Indeed, if $a^* \in \left(0, \frac{1}{c}\right)$, we deduce from (7.61) that $\mathcal{Z}(u_s^{a^*}) < 0$ and therefore (7.73) holds true with

$$\underline{u} := u_s^{a^*}.$$

If instead $a^* \geqslant \frac{1}{c}$, we have that $u_s^a = u_s^{a^*} = 0$, and we deduce from (7.61) and (7.65) that $\mathcal{Z}(u_s^{a^*}) = 0$ and $\mathcal{Z}'(u_s^{a^*}) < 0$, from which (7.73) follows by choosing

$$\underline{u} := u_s^{a^*} + \epsilon$$

with $\epsilon > 0$ sufficiently small.

Now we claim that

$$\mathcal{Z}(u) \leqslant 0 \quad \text{for every } u \in [u_s^{a^*}, u_d^{a^*}]. \quad\quad (7.74)$$

To prove this, in light of (7.73), it suffices to check that $\mathcal{Z}(u) \leqslant 0$ for every $u \in (\underline{u}, u_d^{a^*}]$. Suppose not. Then there exists $u^\sharp \in (\underline{u}, u_d^{a^*}]$ such that $\mathcal{Z}(u) < 0$ for all $[\underline{u}, u^\sharp)$ and $\mathcal{Z}(u^\sharp) = 0$. This gives that $\mathcal{Z}'(u^\sharp) \geqslant 0$. But this inequality is in contradiction with (7.71), and therefore the proof of (7.74) is complete.

The desired claim in (7.57) follows easily from (7.74); hence we focus now on the proof of (7.58).

To this end, we take $\rho > 1$, and we claim that, for every $a^* > 0$ and every $a \in (0, a^*)$,

$$\gamma_a(u_s^{a^*}) \leqslant \gamma_{a^*}(u_s^{a^*})$$

with strict inequality when $a^* \in \left(0, \dfrac{1}{c}\right)$. (7.75)

To prove this, we first notice that if $a < a^* < \frac{1}{c}$, then

$$u_s^{a^*} = \rho c \frac{1 - a^* c}{1 + \rho c} < \rho c \frac{1 - ac}{1 + \rho c} = u_s^a.$$

Hence by (7.46) in Lemma 7.3, we have

$$\gamma_a(u_s^{a^*}) < \frac{u_s^{a^*}}{\rho c} = \gamma_{a^*}(u_s^{a^*}) \quad \text{for } a < a^* < \frac{1}{c},$$

and this establishes (7.75) when $a^* \in \left(0, \frac{1}{c}\right)$. Thus, we now focus on the case $a^* \geqslant \frac{1}{c}$. In this situation, we have that $u_s^{a^*} = 0$ and accordingly $\gamma_a(u_s^{a^*}) = \gamma_a(0) = \gamma_{a^*}(0) = \gamma_{a^*}(u_s^{a^*})$, which completes the proof of (7.75).

In addition, by Propositions 5.1 and 5.7, we have that

$$\gamma_a'(0) = \frac{a}{\rho - 1 + ac} \leqslant \frac{a^*}{\rho - 1 + a^* c} = \gamma_{a^*}'(0) \tag{7.76}$$

for $a \in \left[\dfrac{1}{c}, a^*\right]$.

Moreover, for $u > u_s^a$, if $v = \gamma_a(u)$, we have that $v > \gamma_a(u_s^a) = v_s^a$, thanks to the monotonicity of γ_a, and, as a result,

$$u(1 - u - v - ac) < u(1 - u_s^a - v_s^a - ac) = 0. \tag{7.77}$$

Now we claim that, for all $a < a^*$, $u \in (u_s^{a^*}, u_d^{a^*})$ and $v = \gamma_{a^*}(u)$, we have

$$u(1 - u - v)(a^* - a)(u - \rho c v) < 0. \tag{7.78}$$

Indeed, by the monotonicity of γ_{a^*}, in this situation we have that $v \leqslant \gamma_{a^*}(u_d^{a^*})$, and therefore, by (7.55),

$$1 - u - v > 1 - u_d^{a^*} - \gamma_{a^*}(u_d^{a^*}) = 1 - u_d^{a^*} - 1 + u_d^{a^*} = 0. \tag{7.79}$$

7.3 The Role of the Constant Strategies

Moreover, by (7.47) in Lemma (7.3), we have that $\gamma_{a^*}(u) > \frac{u}{\rho c}$, and hence $u - \rho cv > 0$. Combining this inequality with (7.79), we obtain (7.78), as desired.

Now, by (7.77), for all $a < a^*$, $u \in (u_s^a, u_d^{a^*})$ and $v = \gamma_{a^*}(u)$,

$$0 < -u(1 - u - v - ac) = acu - u(1 - u - v) < a^*cu - u(1 - u - v)$$

and then, by (7.78),

$$\frac{au - \rho v(1 - u - v)}{acu - u(1 - u - v)} - \frac{a^*u - \rho v(1 - u - v)}{a^*cu - u(1 - u - v)}$$
$$= \frac{u(1 - u - v)(a^* - a)(u - \rho cv)}{(acu - u(1 - u - v))(a^*cu - u(1 - u - v))} \quad (7.80)$$
$$< 0.$$

Now we recall the definition of $\mathcal{Z}$ in (7.70), and we claim that

$$\text{if } \check{u} \in (u_s^{a^*}, u_d^{a^*}) \text{ is such that } \mathcal{Z}(\check{u}) = 0, \text{ then } \mathcal{Z}'(\check{u}) < 0. \quad (7.81)$$

To prove this, we let $\check{v} := \gamma_a(\check{u})$, we notice that $\check{v} = \gamma_{a^*}(\check{u})$, and we recall (7.72) and apply it to a trajectory starting at $(\check{u}, \check{v})$, thus finding that

$$\rho\check{v}(1 - \check{u} - v_a(t)) - a\check{u} = \gamma_a'(\check{u})(\check{u}(1 - \check{u} - \check{v}) - ac\check{u}).$$

This and (7.80) yield that

$$0 > \frac{au - \rho v(1 - u - v)}{acu - u(1 - u - v)} - \frac{a^*u - \rho v(1 - u - v)}{a^*cu - u(1 - u - v)}$$
$$= \gamma_a'(\check{u}) - \gamma_{a^*}'(\check{u})$$
$$= \mathcal{Z}'(\check{u}),$$

which proves the desired claim in (7.81).

We now point out that

$$\text{there exists } \underline{u} \in [u_s^{a^*}, u_d^{a^*}] \text{ such that } \mathcal{Z}(\underline{u}) < 0$$
$$\text{and } \mathcal{Z}(u) \leqslant 0 \text{ for every } u \in [u_s^{a^*}, \underline{u}]. \quad (7.82)$$

Indeed, if $a^* \in \left(0, \frac{1}{c}\right)$, this claim follows directly from (7.61) by choosing

$$\underline{u} := u_s^{a^*},$$

while if $a^* \geq \frac{1}{c}$, the claim follows from (7.61) and (7.71) by choosing

$$\underline{u} := u_s^{a^*} + \epsilon$$

with $\epsilon > 0$ sufficiently small.

Now we claim that

$$\mathcal{Z}(u) \leq 0 \qquad \text{for every } u \in [u_s^{a^*}, u_d^{a^*}]. \tag{7.83}$$

Indeed, by (7.82), we know that the claim is true for all $u \in [u_s^{a^*}, \underline{u}]$. Then, the claim for $u \in (\underline{u}, u_d^{a^*}]$ can be proved by contradiction, supposing that there exists $u^\sharp \in (\underline{u}, u_d^{a^*}]$ such that $\mathcal{Z}(u) < 0$ for all $[\underline{u}, u^\sharp)$ and $\mathcal{Z}(u^\sharp) = 0$. This gives that $\mathcal{Z}'(u^\sharp) \geq 0$, which is in contradiction with (7.71).

Having completed the proof of (7.83), one can use it to obtain the desired claim in (7.58). □

Now we perform the proof of Theorem 3.4, analyzing separately the cases $\rho = 1$, $\rho < 1$, and $\rho > 1$.

Proof of Theorem 3.4, case $\rho = 1$ We notice that

$$\mathcal{V}_\mathcal{K} \subseteq \mathcal{V}_\mathcal{A}, \tag{7.84}$$

since $\mathcal{K} \subset \mathcal{A}$.

Also, from Theorem 3.3, part (i), we get that $\mathcal{V}_\mathcal{A} = \mathcal{S}_c$, where $\mathcal{S}_c$ was defined in (6.32). On the other hand, by Lemma 7.2, we know that for $\rho = 1$ and for all $a > 0$ we have $\mathcal{E}(a) = \mathcal{S}_c$. But since every constant a belongs to the set $\mathcal{K}$, we have $\mathcal{E}(a) \subseteq \mathcal{V}_\mathcal{K}$. This shows that $\mathcal{V}_\mathcal{A} = \mathcal{E}(a) \subseteq \mathcal{V}_\mathcal{K}$, and together with (7.84) concludes the proof. □

Proof of Theorem 3.4, Case $\rho < 1$ The situation of this case is sketched in Fig. 7.1. We notice that

$$\mathcal{V}_\mathcal{K} \subseteq \mathcal{V}_\mathcal{A}, \tag{7.85}$$

since $\mathcal{K} \subset \mathcal{A}$. To prove that the inclusion is strict, we aim to find a point $(\bar{u}, \bar{v}) \in \mathcal{V}_\mathcal{A} \setminus \mathcal{V}_\mathcal{K}$. Namely, we have to prove that there exists $(\bar{u}, \bar{v}) \in \mathcal{V}_\mathcal{A}$ such that, for all constant strategies $a > 0$, we have that $(\bar{u}, \bar{v}) \notin \mathcal{E}(a)$, that is, by the characterization in Proposition 5.9, it must hold true that $\bar{v} \geq \gamma_a(\bar{u})$ and $\bar{u} \leq u_\mathcal{M}^a$.

7.3 The Role of the Constant Strategies

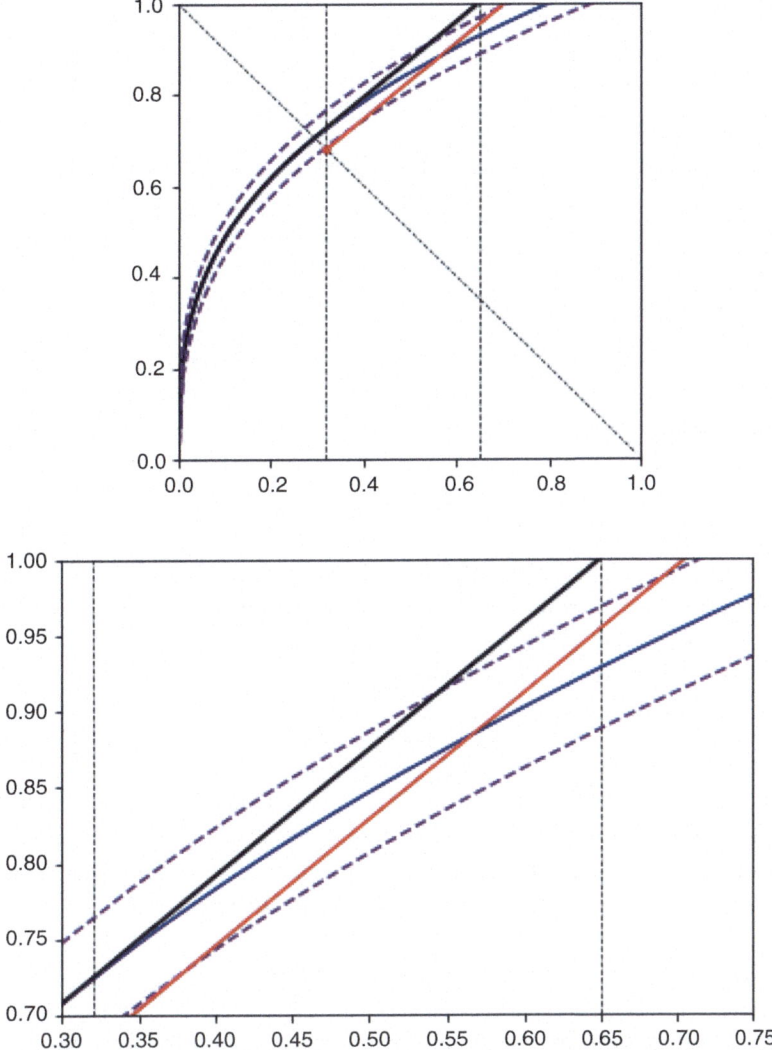

Fig. 7.1 The figures illustrate the functions involved in the proof of Theorem 3.4 for the case $\rho < 1$. The two vertical lines correspond to the values u_d^ε and m. The thick black line represents the boundary of $\mathcal{V}_\mathcal{A}$; the blue line is the graph of $\gamma_0(u)$; the dark violet lines delimit the area where $\gamma_a(u)$ for $a \leqslant \varepsilon$ might be; the red line is the upper limit of $\gamma_a(u)$ for $a > \varepsilon$. The image was realized using a simulation in Python for the values $\rho = 0.35$ and $c = 1.2$

To do this, we define

$$f(u) := \frac{u}{c} + \frac{1-\rho}{1+\rho c} \quad \text{and} \quad m := \min\left\{\frac{\rho c(c+1)}{1+\rho c}, 1\right\}. \tag{7.86}$$

By inspection, one can see that $(u, f(u)) \in [0, 1] \times [0, 1]$ if and only if $u \in [0, m]$. We point out that, by (ii) of Theorem 3.3, for $\rho < 1$ and $u \in [u_s^0, m]$, a point (u, v) belongs to $\mathcal{V}_A$ if and only if $v < f(u)$. Here u_s^0 is defined in (3.8). We underline that the interval $[u_s^0, m]$ is nonempty since

$$u_s^0 = \frac{\rho c}{1+\rho c} < \min\left\{\frac{\rho c(c+1)}{1+\rho c}, 1\right\} = m. \tag{7.87}$$

Now we point out that

$$m \leqslant u_{\mathcal{M}}^a. \tag{7.88}$$

Indeed, by (7.86) we already know that $m \leqslant 1$, and thus if $u_{\mathcal{M}}^a = 1$, the inequality in (7.88) is true. On the other hand, when $u_{\mathcal{M}}^a < 1$, we have that $(u_{\mathcal{M}}^a, 1) \times (0, 1) \subseteq \mathcal{E}(a)$. This and (7.85) give that $(u_{\mathcal{M}}^a, 1) \times (0, 1) \subseteq \mathcal{V}_\mathcal{K} \subseteq \mathcal{V}_A$.

Hence, in view of (3.10), we deduce that

$$\frac{\rho c(c+1)}{1+\rho c} \leqslant u_{\mathcal{M}}^a.$$

In particular, we find that $m \leqslant u_{\mathcal{M}}^a$, and therefore (7.88) is true also in this case.

With this notation, we claim the existence of a value $\bar{v} \in (0, 1]$ such that for all $a > 0$ we have $\gamma_a(m) \leqslant \bar{v} < f(m)$. That is, we prove now that there exists $\theta > 0$ such that

$$\gamma_a(m) + \theta < f(m) \quad \text{for all } a > 0. \tag{7.89}$$

The strategy is to study two cases separately, namely we prove (7.89) for sufficiently small values of a and then for the other values of a.

To prove (7.89) for small values of a, we start by looking at the limit function γ_0 defined in (3.11). One observes that

$$\gamma_0(u_s^0) = v_s^0 = \frac{1}{1+\rho c} = \frac{\rho c}{c(1+\rho c)} + \frac{1-\rho}{1+\rho c} = f(u_s^0). \tag{7.90}$$

Moreover, for all $u \in (u_s^0, m]$, we have that

$$\gamma_0'(u) = \frac{v_s^0}{(u_s^0)^\rho} \rho u^{\rho-1} < \frac{v_s^0}{(u_s^0)^\rho} \rho (u_s^0)^{\rho-1} = \frac{\rho v_s^0}{u_s^0} = \frac{1}{c} = f'(u).$$

7.3 The Role of the Constant Strategies

Hence, using the fundamental theorem of calculus on the continuous functions $\gamma_0(u)$ and $f(u)$, we get

$$\gamma_0(m) = \gamma_0(u_s^0) + \int_{u_s^0}^{m} \gamma_0'(u)\, du$$

$$< f(u_s^0) + \int_{u_s^0}^{m} f'(u)\, du$$

$$= f(m).$$

Then, the quantity

$$\theta_1 := \frac{f(m) - \gamma_0(m)}{4}$$

is positive, and we have

$$\gamma_0(m) + 2\theta_1 < f(m). \tag{7.91}$$

Now, by the uniform convergence of γ_a to γ_0 given by Lemma 6.5, we know that there exists $\varepsilon \in \left(0, \frac{1}{c}\right)$ such that, if $a \in (0, \varepsilon]$,

$$\sup_{u \in [u_s^0, m]} |\gamma_a(u) - \gamma_0(u)| < \theta_1. \tag{7.92}$$

By this and (7.91), we obtain that

$$\gamma_a(m) + \theta_1 < f(m) \quad \text{for all } a \in (0, \varepsilon]. \tag{7.93}$$

We remark that formula (7.93) will give the desired claim in (7.89) for conveniently small values of a.

We are now left with considering the case $a > \varepsilon$. To this end, recalling (3.5), (7.55), by the first statement in Lemma 7.4, used here with $a^* := \varepsilon$, we get

$$\gamma_a(u) \leqslant \gamma_\varepsilon(u) \quad \text{for all } a > \varepsilon \text{ and for all } u \in [u_s^\varepsilon, u_d^\varepsilon]. \tag{7.94}$$

Now we observe that

$$u_d^a \geqslant u_s^\varepsilon. \tag{7.95}$$

Indeed, suppose not, namely

$$u_d^a < u_s^\varepsilon. \tag{7.96}$$

Then, by the monotonicity of γ_a, we have that $\gamma_a(u_d^a) \leqslant \gamma_a(u_s^\varepsilon)$. This and (7.94) yield that $\gamma_a(u_d^a) \leqslant \gamma_\varepsilon(u_s^\varepsilon)$.

Hence, the monotonicity of γ_ε gives that $\gamma_a(u_d^a) \leqslant \gamma_\varepsilon(u_d^a)$. This and (7.55) lead to $1 - u_d^a \leqslant 1 - u_d^\varepsilon$, that is, $u_d^\varepsilon \leqslant u_d^a$. From this inequality, using again (7.96), we deduce that $u_d^\varepsilon < u_s^\varepsilon$. This is in contradiction with (7.59), and thus the proof of (7.95) is complete.

We also notice that

$$u_d^a \geqslant u_d^\varepsilon. \tag{7.97}$$

Indeed, suppose not, say

$$u_d^a < u_d^\varepsilon. \tag{7.98}$$

Then, by (7.95), we have that $u_d^a \in [u_s^\varepsilon, u_d^\varepsilon]$ and therefore we can apply (7.94) to say that $\gamma_a(u_d^a) \leqslant \gamma_\varepsilon(u_d^a)$. Also, by the monotonicity of γ_ε, we have that $\gamma_\varepsilon(u_d^a) \leqslant \gamma_\varepsilon(u_d^\varepsilon)$.

With these items of information and (7.55), we find that

$$1 - u_d^a = \gamma_a(u_d^a) \leqslant \gamma_\varepsilon(u_d^\varepsilon) = 1 - u_d^\varepsilon,$$

and accordingly $u_d^a \geqslant u_d^\varepsilon$. This is in contradiction with (7.98) and establishes (7.97).

Moreover, by (3.5) and (3.8), we know that $u_s^0 > u_s^{a^*}$, for every $a^* > 0$. Therefore, setting

$$\tilde{u}_d^{a^*} := \min\{u_d^{a^*}, u_s^0\},$$

we have that $\tilde{u}_d^{a^*} \in [u_s^{a^*}, u_d^{a^*}]$. Thus, we are in the position of using the first statement in Lemma 7.4 with $a := \varepsilon$ and deduce that

$$\gamma_\varepsilon(\tilde{u}_d^{a^*}) \leqslant \gamma_{a^*}(\tilde{u}_d^{a^*}) \qquad \text{for all} \quad a^* < \varepsilon. \tag{7.99}$$

We also remark that

$$u_d^{a^*} \to u_s^0 \qquad \text{as} \quad a^* \to 0. \tag{7.100}$$

Indeed, up to a subsequence, we can assume that $u_d^{a^*} \to \tilde{u}$ as $a^* \to 0$, for some $\tilde{u} \in [0, 1]$. Also, by (7.55),

$$\gamma_{a^*}(u_d^{a^*}) = 1 - u_d^{a^*},$$

and then the uniform convergence of γ_{a^*} in Lemma 6.5 yields that

$$\gamma_0(\tilde{u}) = 1 - \tilde{u}.$$

7.3 The Role of the Constant Strategies

This and (7.55) lead to $\tilde{u} = u_d^0$. Since

$$u_d^0 = u_s^0 \tag{7.101}$$

in virtue of (3.8), we thus conclude that $\tilde{u} = u_s^0$ and the proof of (7.100) is thereby complete.

As a consequence of (7.100), we have that $\tilde{u}_d^{a^*} \to u_s^0$ as $a^* \to 0$. Hence, using again the uniform convergence of γ_{a^*} in Lemma 6.5, we obtain that $\gamma_{a^*}(\tilde{u}_d^{a^*}) \to \gamma_0(u_s^0)$.

From this and (7.99), we conclude that

$$\gamma_\varepsilon(u_s^0) \leqslant \gamma_0(u_s^0). \tag{7.102}$$

Now we claim that

$$u_d^\varepsilon > u_s^0. \tag{7.103}$$

Indeed, suppose, by contradiction, that

$$u_d^\varepsilon \leqslant u_s^0. \tag{7.104}$$

Then, the monotonicity of γ_ε, together with (7.101) and (7.102), gives that

$$1 - u_d^\varepsilon = \gamma_\varepsilon(u_d^\varepsilon) \leqslant \gamma_\varepsilon(u_s^0) = 1 - u_s^0.$$

From this and (7.104) we deduce that $u_d^\varepsilon = u_s^0$. In particular, we have that $u_s^0 \in (u_s^\varepsilon, u_\mathcal{M}^\varepsilon)$. Accordingly, by (7.45),

$$1 - u_s^0 = 1 - u_d^\varepsilon = \gamma_\varepsilon(u_d^\varepsilon) = \gamma_\varepsilon(u_s^0) < \frac{u_s^0}{\rho c}.$$

As a consequence,

$$u_s^0 > \frac{\rho c}{1 + \rho c},$$

and this is in contradiction with (3.8). The proof of (7.103) is thereby complete.

As a byproduct of (7.101) and (7.103), we have that

$$\begin{aligned}v_d^\varepsilon = \gamma_\varepsilon(u_d^\varepsilon) = 1 - u_d^\varepsilon < 1 - u_s^0 = 1 - u_d^0 \\= \gamma_0(u_d^0) = \gamma_0(u_s^0) = v_s^0.\end{aligned} \tag{7.105}$$

Similarly, by means of (7.97),

$$v_d^a = \gamma_a(u_d^a) = 1 - u_d^a \leqslant 1 - u_d^\varepsilon = \gamma_\varepsilon(u_d^\varepsilon) = v_d^\varepsilon. \qquad (7.106)$$

In light of (7.97), (7.103), (7.105), and (7.106), we can write that

$$1 > u_d^a \geqslant u_d^\varepsilon > u_s^0 > 0 \quad \text{and} \quad 1 > v_s^0 > v_d^\varepsilon \geqslant v_d^a > 0. \qquad (7.107)$$

Now, to complete the proof of (7.89) when $a > \varepsilon$, we consider two cases depending on the order of m and u_d^ε. If $u_d^\varepsilon \geqslant m$, by (7.107), we have that $m < 1$ and $f(m) = 1$. Then,

$$\gamma_a(m) \leqslant \gamma_a(u_d^\varepsilon) \leqslant \gamma_\varepsilon(u_d^\varepsilon) = v_d^\varepsilon < 1 = f(m), \qquad (7.108)$$

thanks to the monotonicity of γ_a, (7.94) and (7.107). We define

$$\theta_2 := \frac{1 - v_d^\varepsilon}{2},$$

which is positive thanks to (7.107). From (7.108), we get that

$$\gamma_a(m) + \theta_2 \leqslant v_d^\varepsilon + \theta_2 < 1 = f(m). \qquad (7.109)$$

This formula proves the claim in (7.89) for $a > \varepsilon$ and $u_d^\varepsilon \geqslant m$.

If instead $u_d^\varepsilon < m$, then we proceed as follows. By (7.107) we have

$$\gamma_a(u_d^a) = v_d^a \leqslant v_d^\varepsilon < v_s^0 = f(u_s^0). \qquad (7.110)$$

Now we set

$$\theta_3 := \frac{f(u_d^\varepsilon) - f(u_s^0)}{2}.$$

Using the definition of f in (7.86), we see that

$$\theta_3 = \frac{u_d^\varepsilon - u_s^0}{2c},$$

and accordingly θ_3 is positive, due to (7.107).

From (7.110) we have

$$\gamma_a(u_d^a) + \theta_3 < f(u_s^0) + \theta_3 < f(u_d^\varepsilon). \qquad (7.111)$$

7.3 The Role of the Constant Strategies

Now we show that, on any trajectory $(u(t), v(t))$ lying on the graph of γ_a, it holds that

$$\dot{v}(t) > \frac{\dot{u}(t)}{c} \quad \text{provided that } u(t) \in (u_d^a, u_{\mathcal{M}}^a). \tag{7.112}$$

To prove this, we first observe that $u(t) > u_d^a > u_s^a$, thanks to (7.59). Hence, we can exploit formula (7.45) of Lemma 7.3 and get that

$$\gamma_a(u(t)) - \frac{u(t)}{\rho c} < 0. \tag{7.113}$$

Also, by the monotonicity of γ_a and (7.55),

$$\gamma_a(u(t)) \geqslant \gamma_a(u_d^a) = 1 - u_d^a > 1 - u(t).$$

From this and (7.113), it follows that

$$\left(\dot{v}(t) - \frac{\dot{u}(t)}{c}\right) = \rho \left(\gamma_a(u(t)) - \frac{u(t)}{\rho c}\right)(1 - u(t) - \gamma_a(u(t))) > 0$$

provided that $u(t) \in (u_d^a, u_{\mathcal{M}}^a)$, and this proves (7.112).

In addition, for such a trajectory $(u(t), v(t))$ we have that

$$\dot{u}(t) = u(t)(1 - u(t) - \gamma_a(u(t)) - ac)$$
$$< u(t)(1 - u(t) - \gamma_a(u_d^a))$$
$$= u(t)(1 - u(t) - 1 + u_d^a)$$
$$< 0,$$

provided that $u(t) \in (u_d^a, u_{\mathcal{M}}^a)$.

From this and (7.112), we get

$$\gamma_a'(u(t)) = \frac{\dot{v}(t)}{\dot{u}(t)} < \frac{1}{c} = f'(u(t)),$$

provided that $u(t) \in (u_d^a, u_{\mathcal{M}}^a)$.

Consequently, taking as initial datum of the trajectory an arbitrary point $(u, \gamma_a(u))$ with $u \in (u_d^a, u_{\mathcal{M}}^a)$, we can write that, for all $u \in (u_d^a, u_{\mathcal{M}}^a)$,

$$\gamma_a'(u) < f'(u).$$

As a result, integrating and using (7.94), for all $u \in (u_d^a, u_{\mathcal{M}}^a)$, we have

$$\gamma_a(u) = \gamma_a(u_d^a) + \int_{u_d^a}^{u} \gamma_a'(u)\,du$$

$$< \gamma_a(u_d^a) + \int_{u_d^a}^{u} f'(u)\,du$$

$$= \gamma_a(u_d^a) + f(u) - f(u_d^a).$$

Then, making use of (7.111), for $u \in (u_d^a, u_{\mathcal{M}}^a)$,

$$\begin{aligned}\gamma_a(u) + \theta_3 &< \gamma_a(u_d^a) + f(u) - f(u_d^a) + \theta_3 \\ &\leqslant f(u) - f(u_d^a) + f(u_d^\varepsilon).\end{aligned} \qquad (7.114)$$

Also, recalling (7.107) and the monotonicity of f, we see that $f(u_d^\varepsilon) \leqslant f(u_d^a)$. Combining this and (7.114), we deduce that

$$\gamma_a(u) + \theta_3 < f(u) \qquad \text{for all } u \in (u_d^a, u_{\mathcal{M}}^a). \qquad (7.115)$$

We also observe that if $u \in (u_d^\varepsilon, u_d^a]$, then the monotonicity of γ_a yields that $\gamma_a(u) \leqslant \gamma_a(u_d^a)$. It follows from this and (7.111) that $\gamma_a(u) + \theta_3 < f(u_d^\varepsilon)$. This and the monotonicity of f give that

$$\gamma_a(u) + \theta_3 < f(u) \qquad \text{for all } u \in (u_d^\varepsilon, u_d^a].$$

Comparing this with (7.115), we obtain

$$\gamma_a(u) + \theta_3 < f(u) \qquad \text{for all } u \in (u_d^\varepsilon, u_{\mathcal{M}}^a)$$

and therefore

$$\gamma_a(u) + \theta_3 \leqslant f(u) \qquad \text{for all } u \in [u_d^\varepsilon, u_{\mathcal{M}}^a]. \qquad (7.116)$$

Now, in view of (7.88), we have that $m \in [u_d^\varepsilon, u_{\mathcal{M}}^a]$. Consequently, we can utilize (7.116) with $u := m$ and find that

$$\gamma_a(m) + \theta_3 \leqslant f(m), \qquad (7.117)$$

which gives (7.89) in the case $a > \varepsilon$ and $u_d^\varepsilon \leqslant m$ (say, in this case with $\theta \leqslant \theta_3/2$).

That is, by (7.93), (7.109), and (7.117) we obtain that (7.89) holds true for

$$\theta := \frac{1}{2} \min\{\theta_1, \theta_2, \theta_3\}.$$

7.3 The Role of the Constant Strategies

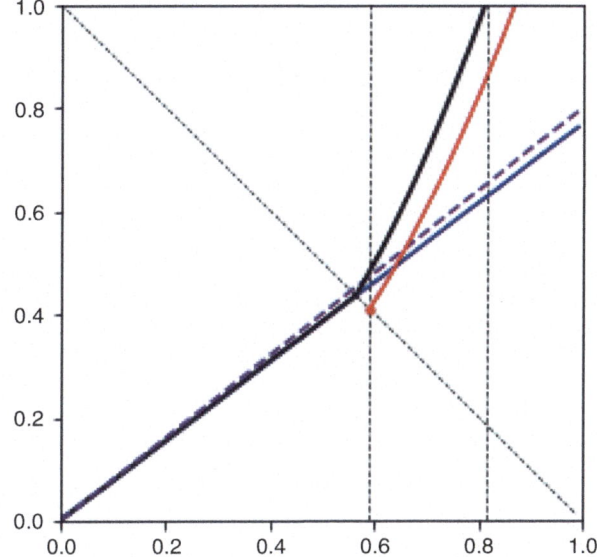

Fig. 7.2 The figure illustrates the functions involved in the proof of Theorem 3.4 for the case $\rho > 1$. The two vertical lines correspond to the values u_d^M and m. The thick black line represents the boundary of $\mathcal{V}_\mathcal{A}$; the blue line is the graph of the line $v = \frac{u}{c}$; the dark violet line is the upper bound for $\gamma_a(u)$ for $a > M$; the red line is $\phi(u)$. The image was realized using a simulation in Python for the values $\rho = 2.3$ and $c = 1.3$.

If we choose $\bar{v} := f(m) - \frac{\theta}{2}$, we have that

$$0 < \gamma_a(m) \leqslant \bar{v} < f(m) \leqslant 1. \tag{7.118}$$

This completes the proof of Theorem 3.4 when $\rho < 1$, in light of the characterizations of $\mathcal{E}(a)$ and $\mathcal{V}_\mathcal{A}$ from Proposition 5.9 and Theorem 3.3, respectively. □

Now we focus on the case $\rho > 1$.

Proof of Theorem 3.4, Case $\rho > 1$ The situation of this case is sketched in Fig. 7.2 As before, the inclusion $\mathcal{V}_\mathcal{K} \subseteq \mathcal{V}_\mathcal{A}$ is trivial since $\mathcal{K} \subset \mathcal{A}$. To prove that it is strict, we aim to find a point $(\bar{u}, \bar{v}) \in \mathcal{V}_\mathcal{A}$ such that $(\bar{u}, \bar{v}) \notin \mathcal{V}_\mathcal{K}$. Thus, we have to prove that there exists $(\bar{u}, \bar{v}) \in \mathcal{V}_\mathcal{A}$ such that, for all constant strategies $a > 0$, we have that $(\bar{u}, \bar{v}) \notin \mathcal{E}(a)$.

To this end, using the characterizations given in Proposition 5.9 and Theorem 3.3, we claim that

there exists a point $(\bar{u}, \bar{v}) \in [0, 1] \times [0, 1]$ satisfying

$$u_\infty \leqslant \bar{u} \leqslant u_\mathcal{M}^a \text{ and } \gamma_a(\bar{u}) \leqslant \bar{v} < \zeta(\bar{u}) \text{ for all } a > 0. \tag{7.119}$$

For this, we let

$$m := \min\left\{1, \frac{c}{(c+1)^{\frac{\rho-1}{\rho}}}\right\}.$$

By (3.13) one sees that

$$u_\infty < m. \tag{7.120}$$

In addition, we point out that

$$m \leqslant u_{\mathcal{M}}^a. \tag{7.121}$$

Indeed, since $m \leqslant 1$, if $u_{\mathcal{M}}^a = 1$, the desired inequality is obvious. If instead $u_{\mathcal{M}}^a < 1$ we have that $(u_{\mathcal{M}}^a, 1) \times (0, 1) \subseteq \mathcal{E}(a) \subseteq \mathcal{V}_\mathcal{K} \subseteq \mathcal{V}_\mathcal{A}$.

Hence, by (3.12), it follows that

$$\frac{c}{(c+1)^{\frac{\rho-1}{\rho}}} \leqslant u_{\mathcal{M}}^a,$$

which leads to (7.121), as desired.

Now we claim that there exists $\theta > 0$ such that

$$\gamma_a(m) + \theta < \zeta(m) \quad \text{for all } a > 0. \tag{7.122}$$

We first show some preliminary facts for $\gamma_a(u)$. For all $a > 0$, we have that $\mathcal{E}(a) \subseteq \mathcal{V}_\mathcal{A}$. Owing to the characterization of $\mathcal{E}(a)$ from Proposition 5.9 and of $\mathcal{V}_\mathcal{A}$ from Theorem 3.3 (which can be used here, thanks to (7.120) and (7.121)), we get that

$$\gamma_a(u) \leqslant \frac{u}{c} \quad \text{for all } u \in (0, u_\infty] \text{ and } a > 0. \tag{7.123}$$

This is true in particular for $u = u_\infty$.

We choose

$$\delta \in \left(0, \frac{\rho-1}{c}\right) \quad \text{and} \quad M := \max\left\{\frac{1}{c}, \frac{\rho + \frac{1}{c} + \delta}{\delta c u_\infty}\right\}, \tag{7.124}$$

and we prove (7.122) by treating separately the cases $a > M$ and $a \in (0, M]$.

We first consider the case $a > M$. We let $(u(t), v(t))$ be a trajectory for (2.1) lying on γ_a, and we show that

$$\dot{v}(t) - \left(\frac{1}{c} + \delta\right)\dot{u}(t) > 0 \tag{7.125}$$

provided that $u(t) > u_\infty$ and $a > M$.

To check this, we observe that

7.3 The Role of the Constant Strategies

$$\dot{v}(t) - \left(\frac{1}{c} + \delta\right)\dot{u}(t)$$
$$= \left[\rho\gamma_a(u(t)) - \left(\frac{1}{c} + \delta\right)u(t)\right](1 - u(t) - \gamma_a(u(t))) + \delta a c u(t)$$
$$\geqslant -\left|\rho + \frac{1}{c} + \delta\right| + \delta a c u_\infty$$
$$> 0,$$

where the last inequality is true thanks to the hypothesis $a > M$ and the definition of M in (7.124). This proves (7.125).

Moreover, for $a > M \geqslant \frac{1}{c}$, we have $\dot{u} < 0$. From this, (7.125) and the invariance of γ_a for the flow, we get

$$\gamma_a'(u(t)) = \frac{\dot{v}(t)}{\dot{u}(t)} < \frac{1}{c} + \delta, \tag{7.126}$$

provided that $u(t) > u_\infty$ and $a > M$.

For this reason and (7.123), we get

$$\gamma_a(u(t)) = \gamma_a(u_\infty) + \int_{u_\infty}^{u(t)} \gamma_a'(\tau)\,d\tau$$
$$\leqslant \frac{u_\infty}{c} + \left(\frac{1}{c} + \delta\right)(u(t) - u_\infty), \tag{7.127}$$

provided that $u(t) > u_\infty$ and $a > M$.

Furthermore, thanks to the choice of δ in (7.124), we have

$$\zeta'(u) = \frac{\rho u^{\rho-1}}{c u_\infty^{\rho-1}} > \frac{\rho}{c} > \frac{1}{c} + \delta \quad \text{for all } u > u_\infty.$$

Since also $\zeta(u_\infty) = \frac{u_\infty}{c}$, by (7.127) we deduce that

$$\gamma_a(u(t)) \leqslant \frac{u_\infty}{c} + \left(\frac{1}{c} + \delta\right)(u(t) - u_\infty)$$
$$< \zeta(u_\infty) + \int_{u_\infty}^{u(t)} \zeta'(\tau)\,d\tau \tag{7.128}$$
$$= \zeta(u(t)),$$

provided that $u(t) > u_\infty$ and $a > M$.

In particular, given any $u > u_\infty$, we can take a trajectory starting at $(u, \gamma_a(u))$ and deduce from (7.128) that

$$\gamma_a(u) \leqslant \frac{u_\infty}{c} + \left(\frac{1}{c} + \delta\right)(u - u_\infty)$$

$$< \zeta(u_\infty) + \int_{u_\infty}^{u} \zeta'(\tau)\,d\tau$$

$$= \zeta(u),$$

whenever $a > M$. We stress that, in light of (7.120), we can take $u := m$ in the above chain of inequalities, concluding that

$$\gamma_a(m) \leqslant \frac{u_\infty}{c} + \left(\frac{1}{c} + \delta\right)(m - u_\infty) < \zeta(m).$$

We rewrite this in the form

$$\gamma_a(m) \leqslant \left(\frac{1}{c} + \delta\right)m - \delta u_\infty < \zeta(m). \tag{7.129}$$

We define

$$\theta_1 := \frac{1}{2}\left[\zeta(m) - \left(\frac{1}{c} + \delta\right)m + \delta u_\infty\right], \tag{7.130}$$

that is positive thanks to the last inequality in (7.129). Then by the first inequality in (7.129) we have

$$\gamma_a(m) + \theta_1 \leqslant \left(\frac{1}{c} + \delta\right)m - \delta u_\infty + \theta_1$$

$$= \frac{1}{2}\left[\left(\frac{1}{c} + \delta\right)m - \delta u_\infty\right] + \frac{\zeta(m)}{2}.$$

Hence, using again the last inequality in (7.129), we obtain that

$$\gamma_a(m) + \theta_1 < \zeta(m), \tag{7.131}$$

which gives the claim in (7.122) for the case $a > M$.

Now we treat the case $a \in (0, M]$. We claim that

$$u_d^M > u_\infty. \tag{7.132}$$

7.3 The Role of the Constant Strategies

Here, we are using the notation u_d^M to denote the point u_d^a when $a := M$. To prove (7.132) we argue as follows. Since $M \geqslant \frac{1}{c}$, by Propositions 5.1 and 5.7 we have

$$\gamma_M'(0) = \frac{M}{\rho - 1 + Mc} < \frac{1}{c}. \tag{7.133}$$

Moreover, since the graph of $\gamma_M(u)$ is a parametrization of a trajectory for (2.1) with $a = M$, we have that $\dot{v}(t) = \gamma_M'(u(t))\dot{u}(t)$. Hence, at all points $(\bar{u}, \bar{v})$ with $\bar{u} \in (0, u_\infty)$ and $\bar{v} = \gamma_M(\bar{u})$ we have

$$\gamma_M'(\bar{u}) = \frac{M\bar{u} - \rho\bar{v}(1 - \bar{u} - \bar{v})}{Mc\bar{u} - \bar{u}(1 - \bar{u} - \bar{v})}. \tag{7.134}$$

We stress that the denominator in the right-hand side of (7.134) is strictly positive, since $M \geqslant \frac{1}{c}$ and $\bar{u} > 0$.

In addition, we have that

$$\frac{1}{c} - \frac{M\bar{u} - \rho\bar{v}(1 - \bar{u} - \bar{v})}{Mc\bar{u} - \bar{u}(1 - \bar{u} - \bar{v})} = \frac{(\rho c\bar{v} - \bar{u})(1 - \bar{u} - \bar{v})}{Mc^2\bar{u} - c\bar{u}(1 - \bar{u} - \bar{v})}. \tag{7.135}$$

Also,

$$u_s^M = 0 < \bar{u} < u_\infty < m \leqslant u_{\mathcal{M}}^M,$$

thanks to (7.120) and (7.121). Hence, we can exploit formula (7.45) in Lemma 7.3 with the strict inequality, thus obtaining that

$$\rho c \bar{v} - \bar{u} = \rho c \gamma_M(\bar{u}) - \bar{u} > 0. \tag{7.136}$$

Moreover, by (7.123),

$$1 - \bar{u} - \bar{v} = 1 - \bar{u} - \gamma_M(\bar{u}) \geqslant 1 - \bar{u} - \frac{\bar{u}}{c} > 1 - u_\infty - \frac{u_\infty}{c} = 0.$$

Therefore, using the latter estimate and (7.136) into (7.135), we get that

$$\frac{1}{c} - \frac{M\bar{u} - \rho\bar{v}(1 - \bar{u} - \bar{v})}{Mc\bar{u} - \bar{u}(1 - \bar{u} - \bar{v})} > 0.$$

From this and (7.134), we have that

$$\gamma_M'(u) < \frac{1}{c} \quad \text{for all } u \in (0, u_\infty).$$

This, together with (7.133) and the fact that $\gamma_M(0) = 0$, gives

$$\gamma_M(u) = \gamma_M(u) - \gamma_M(0) = \int_0^u \gamma_M'(\tau)\, d\tau < \frac{u}{c}$$

for all $u \in (0, u_\infty]$. This inequality yields that

$$\gamma_M(u_\infty) < \frac{u_\infty}{c} = 1 - u_\infty. \tag{7.137}$$

Now, to complete the proof of (7.132), we argue by contradiction and suppose that the claim in (7.132) is false; hence

$$u_d^M \leqslant u_\infty. \tag{7.138}$$

Thus, by (7.137), the monotonicity of $\gamma_M(u)$, and the definition of u_d^M given in (7.55), we get

$$1 - u_d^M = \gamma_M(u_d^M) \leqslant \gamma_M(u_\infty) < 1 - u_\infty,$$

which is in contradiction with (7.138). Hence, (7.132) holds true, as desired.

Also, by the second statement in Lemma 7.4, used here with $a^* := M$,

$$\gamma_a(u) \leqslant \gamma_M(u) \quad \text{for all } u \in [0, u_d^M]. \tag{7.139}$$

We claim that

$$u_d^M \leqslant u_d^a. \tag{7.140}$$

Indeed, suppose, by contradiction, that

$$u_d^M > u_d^a. \tag{7.141}$$

Then, by the monotonicity of γ_a and (7.139), used here with $u := u_d^M$, we find that

$$1 - u_d^a = \gamma_a(u_d^a) \leqslant \gamma_a(u_d^M) \leqslant \gamma_M(u_d^M) = 1 - u_d^M.$$

This entails that $u_d^a \geqslant u_d^M$, which is in contradiction with (7.141), and thus establishes (7.140).

We note in addition that

$$v_d^M = \gamma_M(u_d^M) = 1 - u_d^M < 1 - u_\infty, \tag{7.142}$$

thanks to the definition of (u_d^M, v_d^M) and (7.132).

7.3 The Role of the Constant Strategies

Similarly, by (7.140),

$$v_d^a = \gamma_a(u_d^a) = 1 - u_d^a \leqslant 1 - u_d^M = \gamma_M(u_d^M) = v_d^M. \tag{7.143}$$

Collecting the pieces of information in (7.132), (7.140), (7.142), and (7.143), we thereby conclude that, for all $a \in (0, M]$,

$$\begin{aligned} 0 < u_\infty < u_d^M \leqslant u_d^a < 1 \\ \text{and} \quad 0 < v_d^a \leqslant v_d^M < 1 - u_\infty =: v_\infty < 1. \end{aligned} \tag{7.144}$$

Now we consider two cases depending on the order of m and u_d^M. If $u_d^M \geqslant m$, by (7.144) we have $m < 1$ and $\zeta(m) = 1$. Accordingly, for $a \in (0, M]$, by (7.144) and (7.139) we have

$$\gamma_a(m) \leqslant \gamma_a(u_d^M) \leqslant \gamma_M(u_d^M) = v_d^M < 1 = \zeta(m).$$

Hence, we can define

$$\theta_2 := \frac{1 - v_d^M}{2}$$

and observe that θ_2 is positive by (7.144), thus obtaining that

$$\gamma_a(m) + \theta_2 < \zeta(m). \tag{7.145}$$

This is the desired claim in (7.122) for $a \in (0, M]$ and $u^* \geqslant m$.

If instead $u_d^M < m$, we consider the function

$$\phi(u) := v_d^M \left(\frac{u}{u_d^M} \right)^\rho, \quad \text{for } u \in [u_d^M, m],$$

and we claim that

$$\gamma_a(u) \leqslant \phi(u) \quad \text{for all } a \in (0, M] \text{ and } u \in [u_d^M, m]. \tag{7.146}$$

To prove this, we recall (7.144) and the fact that γ_a is an increasing function to see that

$$\gamma_a(u_d^M) \leqslant \gamma_a(u_d^a) = v_d^a \leqslant v_d^M = \phi(u_d^M). \tag{7.147}$$

Now we remark that

$$\gamma_M(u_d^M) + u_d^M = 1 > 1 - Mc = \gamma_M(u_s^M) + u_s^M,$$

and therefore $u_d^M > u_s^M$. Notice also that $u_d^M < m \leq u_{\mathcal{M}}^M$, thanks to (7.121). As a result, we find that $\rho c \gamma_M(u_d^M) > u_d^M$ by inequality (7.45) in Lemma 7.3. Therefore, if $u \geq u_d^M$ and $v = \phi(u)$, then

$$au\left(1 - \rho c \frac{v_d^M}{(u_d^M)^\rho} u^{\rho-1}\right) = au\left(1 - \frac{\rho c \gamma_M(u_d^M)}{(u_d^M)^\rho} u^{\rho-1}\right)$$

$$< au\left(1 - \left(\frac{u}{u_d^M}\right)^{\rho-1}\right)$$

$$\leq 0$$

$$= \rho\left(v - \frac{v_d^M}{(u_d^M)^\rho} u^\rho\right)(1 - u - v).$$

Using this and (7.77), we deduce that, if $a \in [0, M]$, $u \in [u_d^M, m]$, and $v = \phi(u)$,

$$\frac{au - \rho v(1 - u - v)}{acu - u(1 - u - v)} - \frac{v_d^M}{(u_d^M)^\rho} \rho u^{\rho-1}$$

$$= \frac{au - \rho v(1 - u - v) - (acu - u(1 - u - v))\frac{v_d^M}{(u_d^M)^\rho} \rho u^{\rho-1}}{acu - u(1 - u - v)}$$

$$= \frac{au\left(1 - \rho c \frac{v_d^M}{(u_d^M)^\rho} u^{\rho-1}\right) - \rho(1 - u - v)\left(v - \frac{v_d^M}{(u_d^M)^\rho} u^\rho\right)}{acu - u(1 - u - v)}$$

(7.148)

$$< 0.$$

Now we take $a \in (0, M]$, $u \in [u_d^M, m]$ and suppose that $v = \phi(u) = \gamma_a(u)$, we consider an orbit $(u(t), v(t))$ lying on γ_a with $(u(0), v(0)) = (u, v)$, and we notice that, by (7.77) and (7.148),

$$\gamma_a'(u) = \gamma_a'(u(0))$$

$$= \frac{\dot{v}(0)}{\dot{u}(0)}$$

$$= \frac{au(0) - \rho v(0)(1 - u(0) - v(0))}{acu(0) - u(0)(1 - u(0) - v(0))}$$

$$= \frac{au - \rho v(1 - u - v)}{acu - u(1 - u - v)}$$

(7.149)

$$< \frac{v_d^M}{(u_d^M)^\rho} \rho u^{\rho-1}$$

$$= \phi'(u).$$

7.3 The Role of the Constant Strategies

To complete the proof of (7.146), we define

$$\mathcal{H}(u) := \gamma_a(u) - \phi(u),$$

and we claim that for every $a \in (0, M]$ there exists $\underline{u} \in [u_d^M, m]$ such that

$$\mathcal{H}(\underline{u}) < 0 \text{ and } \mathcal{H}(u) \leqslant 0 \text{ for every } u \in [u_d^M, \underline{u}]. \tag{7.150}$$

Indeed, by (7.147), we know that $\mathcal{H}(u_d^M) \leqslant 0$. Thus, if $\mathcal{H}(u_d^M) < 0$, then we can choose $\underline{u} := u_d^M$ and obtain (7.150). If instead $\mathcal{H}(u_d^M) = 0$, we have that $\gamma_a(u_d^M) = \phi(u_d^M)$ and thus we can exploit (7.149) and find that $\mathcal{H}'(u_d^M) < 0$, from which we obtain (7.150).

Now we claim that, for every $a \in (0, M]$ and $u \in [u_d^M, m]$,

$$\mathcal{H}(u) \leqslant 0. \tag{7.151}$$

For this, given $a \in (0, M]$, we define

$$\mathcal{L} := \{u_* \in [u_d^M, m] \text{ s.t. } \mathcal{H}(u) \leqslant 0 \text{ for every } u \in [u_d^M, u_*]\}$$

and

$$\overline{u} := \sup \mathcal{L}.$$

We remark that $\underline{u} \in \mathcal{L}$, thanks to (7.150), and therefore $\overline{u}$ is well defined. We have that

$$\overline{u} = m, \tag{7.152}$$

otherwise we would have that $\mathcal{H}(\overline{u}) = 0$, and thus $\mathcal{H}'(\overline{u}) < 0$, thanks to (7.149), which would contradict the maximality of $\overline{u}$. Now, the claim in (7.151) plainly follows from (7.152).

We notice that by the inequalities in (7.144) we have

$$\zeta(u) = \frac{v_\infty}{(u_\infty)^\rho} u^\rho > \frac{v_d^M}{(u_d^M)^\rho} u^\rho = \phi(u). \tag{7.153}$$

Then, we define

$$\theta_3 := \frac{\zeta(m) - \phi(m)}{2} \tag{7.154}$$

that is positive thanks to (7.153). We get that

$$\phi(m) + \theta_3 < \zeta(m). \tag{7.155}$$

From this and (7.146), we conclude that

$$\gamma_a(m) + \theta_3 \leqslant \phi(m) + \theta_3 < \zeta(m) \quad \text{for } a \in (0, M]. \tag{7.156}$$

By (7.131), (7.145), and (7.156), we have that (7.122) is true for $\theta = \min\{\theta_1, \theta_2, \theta_3\}$. This also establishes the claim in (7.119), and the proof is completed. □

7.4 The Role of Heaviside Functions

Now, we can complete the proof of Theorem 3.5 by building on the previous work.

Proof of Theorem 3.5 Since the class of Heaviside functions $\mathcal{H}$ is contained in the class of piecewise continuous functions $\mathcal{A}$, we have that

$$\mathcal{V}_\mathcal{H} \subseteq \mathcal{V}_\mathcal{A}, \tag{7.157}$$

and hence we are left with proving the converse inclusion. We treat separately the cases $\rho = 1$, $\rho < 1$, and $\rho > 0$.

If $\rho = 1$, the desired claim follows from Theorem 3.4, part (i).

If $\rho < 1$, we deduce from (3.10) and (7.26) that

$$\mathcal{V}_\mathcal{A} = \mathcal{F}_0 \cup \mathcal{P}, \tag{7.158}$$

where $\mathcal{P}$ has been defined in (7.1) and $\mathcal{F}_0$ in (7.25).

Moreover, by (7.28), we have that

$$\mathcal{F}_0 \subseteq \mathcal{V}_\mathcal{K} \subseteq \mathcal{V}_\mathcal{H}. \tag{7.159}$$

Also, in Proposition 7.1 we construct a Heaviside winning strategy for every point in $\mathcal{P}$. Accordingly, it follows that $\mathcal{P} \subseteq \mathcal{V}_\mathcal{H}$. (7.158) and (7.159) entail that $\mathcal{V}_\mathcal{A} \subseteq \mathcal{V}_\mathcal{H}$, which completes the proof of Theorem 3.5 when $\rho < 1$.

Hence, we now focus on the case $\rho > 1$. By (3.12) and (7.35),

$$\mathcal{V}_\mathcal{A} = \mathcal{S}_c \cup \mathcal{Q}, \tag{7.160}$$

where $\mathcal{S}_c$ was defined in (6.32) and $\mathcal{Q}$ in (7.3).

For every point $(u_0, v_0) \in \mathcal{S}_c$, there exists $\bar{a}$ that is a constant winning strategy for (u_0, v_0), thanks to Proposition 6.4, and therefore $\mathcal{S}_c \subseteq \mathcal{V}_\mathcal{H}$. Moreover, in Proposition 7.1 for every point $(u_0, v_0) \in \mathcal{Q}$, we constructed a Heaviside winning strategy, whence $\mathcal{Q} \subseteq \mathcal{V}_\mathcal{H}$. In light of these observations and (7.160), we see that also in this case $\mathcal{V}_\mathcal{A} \subseteq \mathcal{V}_\mathcal{H}$ and the proof is complete. □

7.5 Pointwise Constraints

This subsection is dedicated to the analysis of $\mathcal{V}_\mathcal{A}$ when we put some constraints on $a(t)$. In particular, we consider $M \geqslant m \geqslant 0$ with $M > 0$ and the set $\mathcal{A}_{m,M}$ of the functions $a(t) \in \mathcal{A}$ with $m \leqslant a(t) \leqslant M$ for all $t > 0$. We will prove Theorem 3.6 via a technical proposition giving informative bounds on $\mathcal{V}_{m,M}$.

For this, we denote by (u_s^m, v_s^m) the point (u_s, v_s) introduced in (3.5) when $a(t) = m$ for all $t > 0$ (this when $mc < 1$, and we use the convention that $(u_s^m, v_s^m) = (0, 0)$ when $mc \geqslant 1$). In this setting, we have the following result obtaining explicit bounds on the favorable set $\mathcal{V}_{m,M}$:

Proposition 7.5 *Let $M \geqslant m \geqslant 0$ with $M > 0$ and*

$$\varepsilon \in \left(0, \min\left\{\frac{M(c+1)}{M+1}, 1\right\}\right). \tag{7.161}$$

Then:

(i) If $\rho < 1$, we have

$$\mathcal{V}_{m,M} \subseteq \left\{(u, v) \in [0, 1] \times [0, 1] \text{ s.t. } v < f_\varepsilon(u)\right\}, \tag{7.162}$$

where $f_\varepsilon : [0, u_\mathcal{M}] \to [0, 1]$ is the continuous function given by

$$f_\varepsilon(u) = \begin{cases} \dfrac{(u_s^m)^{1-\rho} u^\rho}{\rho c} & \text{if } u \in [0, u_s^m), \\ \dfrac{u}{\rho c} & \text{if } u \in [u_s^m, u_s^0), \\ \dfrac{u}{c} + \dfrac{1-\rho}{1+\rho c} & \text{if } u \in [u_s^0, u_1), \\ hu + p & \text{if } u \in [u_1, 1], \end{cases}$$

with the convention that the first interval is empty if $m \geqslant \frac{1}{c}$, the second interval is empty if $m = 0$, and h, u_1 and p take the following values:

$$h := \frac{1}{c}\left(1 - \frac{\varepsilon^2(1-\rho)}{M(1+\rho c)(c+1-\varepsilon)^2 + \varepsilon(\rho c + \rho + \varepsilon - \varepsilon \rho)}\right),$$

$$u_1 := \frac{c(\rho c + \rho + \varepsilon - \varepsilon \rho)}{(1+\rho c)(c+1-\varepsilon)}$$

and

$$p := \frac{c+1-hc(\rho c+\rho+\varepsilon-\varepsilon\rho)}{(1+\rho c)(c+1-\varepsilon)}.$$

(ii) If $\rho > 1$, we have

$$\mathcal{V}_{m,M} \subseteq \Big\{(u,v) \in [0,1]\times[0,1] \text{ s.t. } v < g_\varepsilon(u)\Big\},$$

where $g_\varepsilon : [0, u_\mathcal{M}] \to [0,1]$ is the continuous function given by

$$g_\varepsilon(u) = \begin{cases} ku & \text{if } u \in [0, u_2), \\ \dfrac{u}{c} + q & \text{if } u \in [u_2, u_3), \\ \dfrac{(1-u_3)u^\rho}{(u_3)^\rho} & \text{if } u \in [u_3, 1] \end{cases}$$

for the following values:

$$k := \frac{(c+1-\varepsilon)M}{(\rho-1)\varepsilon c + (c+1-\varepsilon)Mc},$$

$$q := \frac{(kc-1)(1-\varepsilon)}{c(k-k\varepsilon+1)},$$

$$u_2 := \frac{1-\varepsilon}{k-k\varepsilon+1}$$

and
$$u_3 := \frac{c+1-\varepsilon}{(c+1)(k-k\varepsilon+1)}.$$

We observe that it might be that for some $u \in [0,1]$ we have $f_\varepsilon(u) > 1$ or $g_\varepsilon(u) > 1$. In this case, Proposition 7.5 would produce the trivial result that $\mathcal{V}_{m,M} \cap (\{u\} \times [0,1]) \subseteq \{u\} \times [0,1]$.

On the other hand, a suitable choice of ε would lead to nontrivial consequences entailing, in particular, the proof of Theorem 3.6.

Proof of Proposition 7.5 We start by proving the claim in (i). For this, we will show that

$$\mathcal{V}_{m,M} \subset \mathcal{D} := \Big\{(u,v) \in [0,1]\times[0,1] \text{ s.t. } v < f_\varepsilon(u)\Big\}. \tag{7.163}$$

We remark that once (7.163) is established, then the desired claim in (7.162) plainly follows by taking the complement sets.

To prove (7.163) we first show that

$$0 \leqslant u_s^m < u_s^0 < u_1 < 1. \tag{7.164}$$

7.5 Pointwise Constraints

Notice, as a byproduct, that the above inequalities also give that f_ε is well defined. To prove (7.164) we notice that, by (3.5), (3.8), and (7.43),

$$0 \leqslant u_s^m = \max\left\{0, \frac{1-mc}{1+\rho c}\rho c\right\} < \frac{\rho c}{1+\rho c} = u_s^0.$$

Actually the first inequality is strict if $m < \frac{1}{c}$.

Next, one can check that, since $\varepsilon > 0$,

$$u_s^0 - u_1 = \frac{\rho c}{1+\rho c} - \frac{c(\rho c + \rho + \varepsilon - \varepsilon\rho)}{(1+\rho c)(c+1-\varepsilon)} = -\frac{c\varepsilon}{(1+\rho c)(c+1-\varepsilon)} < 0.$$

Furthermore, since $\varepsilon < 1$,

$$u_1 - 1 = \frac{c(\rho c + \rho + \varepsilon - \varepsilon\rho)}{(1+\rho c)(c+1-\varepsilon)} - 1 = \frac{(\varepsilon-1)(c+1)}{(1+\rho c)(c+1-\varepsilon)} < 0.$$

These observations prove (7.164), as desired.

Now we point out that

$$f_\varepsilon \text{ is a continuous function.} \tag{7.165}$$

Indeed,

$$\frac{(u_s^m)^{1-\rho}}{\rho c}(u_s^m)^\rho = \frac{u_s^m}{\rho c} \quad \text{and} \quad \frac{u_s^0}{\rho c} = \frac{u_s^0}{c} + \frac{1-\rho}{1+\rho c}. \tag{7.166}$$

Furthermore, by the definitions of p and u_1 we see that

$$\begin{aligned} p &= \frac{c+1}{(1+\rho c)(c+1-\varepsilon)} - \frac{hc(\rho c + \rho + \varepsilon - \varepsilon\rho)}{(1+\rho c)(c+1-\varepsilon)} \\ &= \frac{c+1}{(1+\rho c)(c+1-\varepsilon)} - hu_1. \end{aligned} \tag{7.167}$$

Moreover, from the definition of u_1,

$$\frac{u_1}{c} + \frac{1-\rho}{1+\rho c} = \frac{c+1}{(1+\rho c)(c+1-\varepsilon)}.$$

Combining this and (7.167), we deduce that

$$\frac{u_1}{c} + \frac{1-\rho}{1+\rho c} = hu_1 + p. \tag{7.168}$$

This observation and (7.166) entail the desired claim in (7.165).

Next, we show that

$$f_\varepsilon(u) > 0 \quad \text{for } u > 0. \tag{7.169}$$

To prove this, we note that for $u \in (0, u_s^m)$ the function $f_\varepsilon(u)$ is an exponential times the positive constant

$$\frac{(u_s^m)^{1-\rho}}{\rho c},$$

and hence it is positive.

If $u \in [u_s^m, u_s^0)$, then $f_\varepsilon(u)$ is a linear function and it is positive since $\rho c > 0$.

On $[u_s^0, u_1)$, $f_\varepsilon(u)$ coincides with a linear function with positive angular coefficient, and hence we have

$$f_\varepsilon(u) \geqslant \min_{u \in [u_s^0, u_1)} f_\varepsilon(u) = f_\varepsilon(u_s^0) = \frac{u_s^0}{\rho c} > 0.$$

By inspection one can check that $h > 0$.

Therefore, in the interval $[u_1, 1]$ we have

$$f_\varepsilon(u) \geqslant \min_{u \in [u_1, 1]} f_\varepsilon(u) = f_\varepsilon(u_1) \geqslant \frac{u_s^0}{\rho c} > 0.$$

This completes the proof of (7.169).

Let us notice that, as a consequence of (7.169),

$$((0, 1] \times \{0\}) \subset \mathcal{D}. \tag{7.170}$$

Now we show that

$$\text{for any strategy } a \in \mathcal{A}_{m,M}, \text{ no trajectory enters } \mathcal{D}. \tag{7.171}$$

To apply Lemma (4.6), we compute the velocity of a trajectory in the outward normal direction at $\partial \mathcal{D} \cap \{v = f_\varepsilon(u)\}$.

For every $u \in [0, u_s^m)$ we have that this normal velocity is

$$\dot{v} - \frac{(u_s^m)^{1-\rho} \rho(u)^{\rho-1} \dot{u}}{\rho c}$$

$$= \rho \left(v - \frac{(u_s^m)^{1-\rho} u^\rho}{\rho c} \right)(1 - u - v) - au \left(1 - \frac{(u_s^m)^{1-\rho}}{u^{1-\rho}} \right). \tag{7.172}$$

7.5 Pointwise Constraints

Notice that the term

$$v - \frac{(u_s^m)^{1-\rho} u^\rho}{\rho c}$$

vanishes on $\partial \mathcal{D} \cap \{v = f_\varepsilon(u)\}$ when $u \in [0, u_s^m]$.

Also, for all $u \in [0, u_s^m]$ we have

$$1 - \frac{(u_s^m)^{1-\rho}}{u^{1-\rho}} \leqslant 0,$$

and thus the left-hand side in (7.172) is nonnegative with equality only in u_s^m. So (4.2) is satisfied for $u \in (0, u_s^m)$, and (4.3) is satisfied for $u = 0$ and $u = u_s^m$.

It remains to verify the hypothesis at points of $\partial \mathcal{D} \cap \{v = f_\varepsilon(u)\}$ with $u \in [u_s^m, 1]$. We first consider this type of points when $[u_s^m, u_s^0]$. At these points, we have that the velocity in the outward normal direction on $\{v = \frac{u}{\rho c}\}$ is

$$\dot{v} - \frac{\dot{u}}{\rho c} = \left(\rho v - \frac{u}{\rho c}\right)(1 - u - v) + au\left(\frac{1}{\rho} - 1\right).$$

Expressing u with respect to v on $\partial \mathcal{D} \cap \{v = f_\varepsilon(u)\}$ with $u \in [u_s^m, u_s^0]$, we have

$$\begin{aligned}\dot{v} - \frac{\dot{u}}{\rho c} &= v(\rho - 1)(1 - \rho c v - v) + a\rho c v \frac{1 - \rho}{\rho} \\ &= v(1 - \rho)(\rho c v + v - 1 + ac).\end{aligned} \quad (7.173)$$

We also remark that, for these points,

$$v \geqslant v_s^m = \frac{1 - mc}{1 + \rho c} \geqslant \frac{1 - ac}{1 + \rho c},$$

thanks to (3.5). This gives that the quantity in (7.173) is strictly positive for $u \in (u_s^m, u_s^0)$ and null for $u = u_s^m$ and $u = u_s^0$, and as a consequence, we have proved (4.2) for $u \in (u_s^m, u_s^0)$ and (4.3) for $u = u_s^m$ and $u = u_s^0$.

It remains to consider the case $u \in [u_s^0, 1]$; we now consider the interval $u \in [u_s^0, u_1]$. In this interval, the boundary $\partial \mathcal{D} \cap \{v = f_\varepsilon(u)\}$ lies on the line $v = \frac{u}{c} + \frac{1-\rho}{1+\rho c}$. The velocity of a trajectory starting at a point $(u, v) \in \partial \mathcal{D} \cap \{v = f_\varepsilon(u)\}$ in the outward normal direction with respect to $\partial \mathcal{D}$ is given by

$$\dot{v} - \frac{1}{c}\dot{u} = \left(\rho v - \frac{u}{c}\right)(1 - u - v). \quad (7.174)$$

We also observe that, in light of (3.8),

$$u \geqslant u_s^0 = \frac{\rho c}{1 + \rho c},$$

and therefore, for any $u \in [u_s^0, u_1]$ lying on the above line,

$$1 - u - v = 1 - u - \frac{u}{c} - \frac{1 - \rho}{1 + \rho c} = (c + 1)\left(\frac{\rho}{1 + \rho c} - \frac{u}{c}\right) \leqslant 0$$

and

$$\rho v - \frac{u}{c} = \frac{\rho u}{c} + \frac{\rho(1 - \rho)}{1 + \rho c} - \frac{u}{c} = (1 - \rho)\left(\frac{\rho}{1 + \rho c} - \frac{u}{c}\right) \leqslant 0,$$

with equality in u_s^0. Using these pieces of information in (7.174), we conclude that (4.2) for a point $(u, v) \in \partial \mathcal{D} \cap \{v = f_\varepsilon(u)\}$ is satisfied for $u \in (u_s^0, u_1]$ and (4.3) is satisfied for $u = u_s^0$. We need to verify the case $u \in [u_1, 1]$.

We consider now the interval $[u_1, 1]$. In this interval, the component of the velocity of a trajectory at a point on the straight line given by $v = hu + p$ in the orthogonal outward pointing direction is

$$(\dot{u}, \dot{v}) \cdot \frac{(-h, 1)}{\sqrt{1 + h^2}} = \frac{(\rho v - hu)(1 - u - v) - au(1 - hc)}{\sqrt{1 + h^2}}$$
$$= \frac{((1 - \rho)hu - \rho p)(u + v - 1) - au(1 - hc)}{\sqrt{1 + h^2}}.$$
(7.175)

We observe that, if $u \in [u_1, 1]$,

$$(1 - \rho)hu - \rho p \geqslant (1 - \rho)hu_1 - \rho p$$
$$= hu_1 - \rho(hu_1 + p)$$
$$= hu_1 - \rho\left(\frac{u_1}{c} + \frac{1 - \rho}{1 + \rho c}\right)$$
$$= hu_1 - \rho\left(\frac{\rho c + \rho + \varepsilon - \varepsilon \rho}{(1 + \rho c)(c + 1 - \varepsilon)} + \frac{1 - \rho}{1 + \rho c}\right)$$
$$= hu_1 - \frac{\rho(c + 1)}{(1 + \rho c)(c + 1 - \varepsilon)},$$
(7.176)

thanks to (7.168).

7.5 Pointwise Constraints

We also remark that

$$hu_1$$
$$= \left(1 - \frac{\varepsilon^2(1-\rho)}{M(1+\rho c)(c+1-\varepsilon)^2 + \varepsilon(\rho c + \rho + \varepsilon - \varepsilon\rho)}\right) \frac{\rho c + \rho + \varepsilon - \varepsilon\rho}{(1+\rho c)(c+1-\varepsilon)}$$
$$= \frac{\rho c + \rho + \varepsilon - \varepsilon\rho}{(1+\rho c)(c+1-\varepsilon)}$$
$$- \frac{\varepsilon^2(1-\rho)(\rho c + \rho + \varepsilon - \varepsilon\rho)}{(M(1+\rho c)(c+1-\varepsilon)^2 + \varepsilon(\rho c + \rho + \varepsilon - \varepsilon\rho))(1+\rho c)(c+1-\varepsilon)}.$$

Accordingly,

$$hu_1 - \frac{\rho(c+1)}{(1+\rho c)(c+1-\varepsilon)}$$
$$= \frac{\varepsilon(1-\rho)}{(1+\rho c)(c+1-\varepsilon)}$$
$$- \frac{\varepsilon^2(1-\rho)(\rho c + \rho + \varepsilon - \varepsilon\rho)}{(M(1+\rho c)(c+1-\varepsilon)^2 + \varepsilon(\rho c + \rho + \varepsilon - \varepsilon\rho))(1+\rho c)(c+1-\varepsilon)}$$
$$= \frac{\varepsilon(1-\rho)}{(1+\rho c)(c+1-\varepsilon)} \left(1 - \frac{\varepsilon(\rho c + \rho + \varepsilon - \varepsilon\rho)}{M(1+\rho c)(c+1-\varepsilon)^2 + \varepsilon(\rho c + \rho + \varepsilon - \varepsilon\rho)}\right)$$
$$= \frac{\varepsilon(1-\rho)}{(1+\rho c)(c+1-\varepsilon)} \cdot \frac{M(1+\rho c)(c+1-\varepsilon)^2}{M(1+\rho c)(c+1-\varepsilon)^2 + \varepsilon(\rho c + \rho + \varepsilon - \varepsilon\rho)}$$
$$= \frac{\varepsilon M(1-\rho)(c+1-\varepsilon)}{M(1+\rho c)(c+1-\varepsilon)^2 + \varepsilon(\rho c + \rho + \varepsilon - \varepsilon\rho)}.$$

From this and (7.176), we gather that

$$(1-\rho)hu - \rho p$$
$$\geqslant \frac{\varepsilon M(1-\rho)(c+1-\varepsilon)}{M(1+\rho c)(c+1-\varepsilon)^2 + \varepsilon(\rho c + \rho + \varepsilon - \varepsilon\rho)}. \tag{7.177}$$

Furthermore, we point out that, when $[u_1, 1)$ and $v = hu + p$,

$$u + v - 1 \geqslant u_1 + hu_1 + p - 1$$
$$= u_1 + \frac{u_1}{c} + \frac{1-\rho}{1+\rho c} - 1$$
$$= \frac{(c+1)(\rho c + \rho + \varepsilon - \varepsilon\rho)}{(1+\rho c)(c+1-\varepsilon)} - \frac{\rho(c+1)}{1+\rho c}$$
$$= \frac{\varepsilon(c+1)}{(1+\rho c)(c+1-\varepsilon)}$$
$$> \frac{\varepsilon}{c+1-\varepsilon},$$

thanks to (7.168).

Combining this inequality and (7.177), we deduce that

$$((1-\rho)hu - \rho p)(u+v-1)$$
$$> \frac{\varepsilon^2 M(1-\rho)}{M(1+\rho c)(c+1-\varepsilon)^2 + \varepsilon(\rho c + \rho + \varepsilon - \varepsilon\rho)}.$$

Therefore, noticing that $h < \frac{1}{c}$,

$$((1-\rho)hu - \rho p)(u+v-1) - au(1-hc)$$
$$> \frac{\varepsilon^2 M(1-\rho)}{M(1+\rho c)(c+1-\varepsilon)^2 + \varepsilon(\rho c + \rho + \varepsilon - \varepsilon\rho)} - Mu(1-hc)$$
$$= \frac{\varepsilon^2 M(1-\rho)(1-u)}{M(1+\rho c)(c+1-\varepsilon)^2 + \varepsilon(\rho c + \rho + \varepsilon - \varepsilon\rho)},$$

which is strictly positive.

Using this information in (7.175), we can thereby conclude that (4.2) is verified for $(u, f_\varepsilon(u)) \in \partial \mathcal{D}$ with $u \in [u_1, 1)$.

In this way, we have shown that either (4.2) or (4.3) holds for $(u, f_\varepsilon(u)) \in \partial \mathcal{D}$ for $u \in [0, 1]$ and (4.3) holds in a finite number of points, so we can apply Lemma 4.6. Hence, no trajectory can enter $\mathcal{D}$, and the proof of (7.171) is complete.

By (7.170) and (7.171), no trajectory starting outside $\mathcal{D}$ can arrive in $(0, 1] \times [0, 1]$ when the bound $m \leqslant a(t) \leqslant M$ holds, and hence (7.163) is true. Therefore the statement (i) in Proposition 7.5 is true.

Now we establish the claim in (ii). To this end, we point out that claim (ii) is equivalent to

$$\mathcal{V}_{m,M} \subseteq \mathcal{G} := \left\{ (u, v) \in [0, 1] \times [0, 1] \text{ s.t. } v < g_\varepsilon(u) \right\}. \tag{7.178}$$

7.5 Pointwise Constraints

First, we point out that

$$g_\varepsilon \text{ is a well-defined continuous function.} \tag{7.179}$$

Indeed, one can easily check for $\varepsilon \in (0, 1)$ that

$$\begin{aligned} 0 &< u_2 \\ &= \frac{1-\varepsilon}{k-k\varepsilon+1} - \frac{c+1-\varepsilon}{(c+1)(k-k\varepsilon+1)} + u_3 \\ &= -\frac{c\varepsilon}{(c+1)(k-k\varepsilon+1)} + u_3 \\ &< u_3 \\ &< \frac{c+1}{(c+1)(k-k\varepsilon+1)} \\ &< 1. \end{aligned} \tag{7.180}$$

Then, one checks that

$$ku_2 = \frac{u_2}{c} + q,$$

and hence g_ε is continuous at the point u_2. In addition, one can check that g_ε is continuous at the point u_3 by observing that

$$\begin{aligned} \frac{u_3}{c} + q - (1 - u_3) &= \frac{(c+1)u_3}{c} + q - 1 \\ &= \frac{c+1-\varepsilon}{c(k-k\varepsilon+1)} + \frac{(kc-1)(1-\varepsilon)}{c(k-k\varepsilon+1)} - 1 \\ &= \frac{c+1-\varepsilon + (kc-1)(1-\varepsilon) - c(k-k\varepsilon+1)}{c(k-k\varepsilon+1)} = 0. \end{aligned} \tag{7.181}$$

This completes the proof of (7.179).

Now we show that

$$g_\varepsilon(u) > 0 \quad \text{for every } u \in (0, 1]. \tag{7.182}$$

We have that $k > 0$ for every $\varepsilon < 1$, and therefore $g_\varepsilon(u) > 0$ for all $u \in (0, u_2)$. Also, since $g_\varepsilon(u_2) = ku_2 > 0$ and g_ε is linear in (u_2, u_3), we have that $g_\varepsilon(u) > 0$ for all $u \in (u_2, u_3)$.

Moreover, in the interval $\in [u_3, 1]$, we have that g_ε is an exponential function multiplied by a positive constant, thanks to (7.180); hence it is positive. These considerations prove (7.182).

As a consequence of (7.182), we have that

$$((0, 1] \times \{0\}) \subset \mathcal{G}. \tag{7.183}$$

Now we claim that

$$\text{for any strategy } a \in \mathcal{A}_{m,M}, \text{ no trajectory enters } \mathcal{G}. \tag{7.184}$$

To prove (7.184), we want to apply Lemma 4.6. We do this by showing that the outward pointing derivative of the trajectory is positive up to a finite number of points, where it is zero, according to the computation below.

At a point on the line $v = ku$, the velocity of a trajectory in the direction that is orthogonal to $\partial \mathcal{G}$ for $u \in (0, u_2]$ and pointing outward is

$$(\dot{u}, \dot{v}) \cdot \frac{(-k, 1)}{\sqrt{1+k^2}} = \frac{(\rho v - ku)(1 - u - v) - au(1 - kc)}{\sqrt{1+k^2}}. \tag{7.185}$$

We also note that

$$kc = \frac{(c + 1 - \varepsilon)M}{(\rho - 1)\varepsilon + (c + 1 - \varepsilon)M} < 1, \tag{7.186}$$

and therefore, at a point on $v = ku$ with $u \in (0, u_2]$,

$$1 - u - v \geq 1 - u_2 - ku_2$$
$$= 1 - \frac{(1+k)(1-\varepsilon)}{k - k\varepsilon + 1}$$
$$= \frac{\varepsilon}{k(1-\varepsilon) + 1}$$
$$= \frac{\varepsilon c}{kc(1-\varepsilon) + c}$$
$$> \frac{\varepsilon c}{1 + c - \varepsilon}.$$

This inequality entails that

$$k = \frac{(1 + c - \varepsilon)M}{(\rho - 1)\varepsilon c + (1 + c - \varepsilon)Mc}$$
$$= \frac{M}{\frac{(\rho-1)\varepsilon c}{1+c-\varepsilon} + Mc}$$
$$> \frac{M}{(\rho - 1)(1 - u - v) + Mc}.$$

7.5 Pointwise Constraints

Consequently,

$$(\rho - 1)(1 - u - v)k > M(1 - kc).$$

From this and (7.185), one deduces that, for all $u \in (0, u_2]$, $a \leqslant M$, and $v = ku$,

$$\begin{aligned}
(\dot{u}, \dot{v}) \cdot \frac{(-k, 1)}{\sqrt{1+k^2}} &= \frac{ku(\rho - 1)(1 - u - v) - au(1 - kc)}{\sqrt{1+k^2}} \\
&> \frac{Mu(1 - kc) - au(1 - kc)}{\sqrt{1+k^2}} \\
&> 0,
\end{aligned}$$

thus satisfying (4.2). Moreover, since $(0, 0)$ is an equilibrium, (4.3) holds for $u = 0$.

It remains to consider the portions of $\partial \mathcal{G} \cap ((0, 1) \times (0, 1))$ given by

$$\left\{ u \in [u_2, u_3) \text{ and } v = \frac{u}{c} + q \right\} \tag{7.187}$$

and by

$$\left\{ u \in [u_3, 1] \text{ and } v = \frac{(1 - u_3)u^\rho}{(u_3)^\rho} \right\}. \tag{7.188}$$

Let us deal with the case in (7.187). In this case, the velocity of a trajectory in the direction orthogonal to $\partial \mathcal{G}$ for $u \in [u_2, u_3)$ and pointing outward is

$$(\dot{u}, \dot{v}) \cdot \frac{(-1, c)}{\sqrt{1+c^2}} = \frac{(\rho c v - u)(1 - u - v)}{\sqrt{1+c^2}}. \tag{7.189}$$

Recalling (7.161), we also observe that

$$\begin{aligned}
k - \frac{1}{\rho c} &= \frac{1}{c} \left(\frac{(c + 1 - \varepsilon)M}{(\rho - 1)\varepsilon + (c + 1 - \varepsilon)M} - \frac{1}{\rho} \right) \\
&= \frac{(\rho - 1)\big((c + 1 - \varepsilon)M - \varepsilon\big)}{\rho c\big((\rho - 1)\varepsilon + (c + 1 - \varepsilon)M\big)} \\
&> 0.
\end{aligned} \tag{7.190}$$

Thus, on the line given by $v = \frac{u}{c} + q$ we have that

$$\rho c v - u = (\rho - 1)u + \rho c q$$
$$\geqslant (\rho - 1)u_2 + \rho c q$$
$$= \frac{(\rho - 1)(1 - \varepsilon)}{k - k\varepsilon + 1} + \frac{\rho(kc - 1)(1 - \varepsilon)}{k - k\varepsilon + 1}$$
$$= (1 - \varepsilon)\frac{(\rho - 1) + \rho(kc - 1)}{k - k\varepsilon + 1} \quad (7.191)$$
$$= \frac{(1 - \varepsilon)(\rho kc - 1)}{k - k\varepsilon + 1}$$
$$> 0,$$

where (7.190) has been used in the latter inequality.

In addition, recalling (7.181),

$$1 - u - v > 1 - u_3 - \frac{u_3}{c} - q = 1 - u_3 - 1 + u_3 = 0.$$

From this and (7.191), we gather that the velocity calculated in (7.189) is positive in $[u_2, u_3)$ (satisfying (4.2)) and null in u_3 (satisfying (4.3)).

Next, we focus on the portion of the boundary described in (7.188) by considering $u \in [u_3, 1]$. That is, we now compute the component of the velocity at a point on $\partial \mathcal{G}$ for $u \in [u_3, 1]$ in the direction that is orthogonal to $\partial \mathcal{G}$ and pointing outward, that is,

$$(\dot{u}, \dot{v}) \cdot \frac{(-\rho \frac{1-u_3}{(u_3)^\rho} u^{\rho-1}, 1)}{\sqrt{1 + \rho^2 \frac{(1-u_3)^2}{(u_3)^{2\rho}} u^{2\rho-2}}}$$

$$= \frac{\rho(1 - u - v)\left(v - \frac{1-u_3}{(u_3)^\rho} u^\rho\right) - au\left(1 - \rho c \frac{1-u_3}{(u_3)^\rho} u^{\rho-1}\right)}{\sqrt{1 + \rho^2 \frac{(1-u_3)^2}{(u_3)^{2\rho}} u^{2\rho-2}}}$$

$$= \frac{au\left(\rho c \frac{1-u_3}{(u_3)^\rho} u^{\rho-1} - 1\right)}{\sqrt{1 + \rho^2 \frac{(1-u_3)^2}{(u_3)^{2\rho}} u^{2\rho-2}}} \quad (7.192)$$

$$\geqslant \frac{au\left(\rho c \frac{1-u_3}{u_3} - 1\right)}{\sqrt{1 + \rho^2 \frac{(1-u_3)^2}{(u_3)^{2\rho}} u^{2\rho-2}}}.$$

Now we notice that

$$\rho c(1 - u_3) = \rho c\left(\frac{u_3}{c} + q\right)$$
$$= \rho u_3 + \rho c q$$

7.5 Pointwise Constraints

$$= \rho u_3 + \frac{\rho(kc - 1)(1 - \varepsilon)(c + 1)u_3}{c + 1 - \varepsilon},$$

thanks to (7.181).

As a result, using (7.190),

$$\begin{aligned}
\rho c(1 - u_3) &> \rho u_3 + \frac{(1 - \rho)(1 - \varepsilon)(c + 1)u_3}{c + 1 - \varepsilon} \\
&= \frac{u_3}{c + 1 - \varepsilon}\Big(\rho(c + 1 - \varepsilon) + (1 - \rho)(1 - \varepsilon)(c + 1)\Big) \\
&= \frac{u_3\big((1 - \varepsilon)(c + 1) + \varepsilon\rho c\big)}{c + 1 - \varepsilon} \\
&= u_3 + \frac{\varepsilon c u_3(\rho - 1)}{c + 1 - \varepsilon} \\
&> u_3.
\end{aligned}$$

This gives that the quantity in (7.192) is positive, proving (4.2) for $u \in [u_3, 1]$.

Hence, since (4.3) is verified at $u = 0, u_3$ and (4.2) is verified for $u \in [0, 1] \setminus \{0, u_3\}$, we can apply Lemma 4.6 and get (7.184).

Since no trajectory can enter $\mathcal{G}$ for any a with $m \leqslant a \leqslant M$, we get that no point $(u, v) \in \mathcal{G}^c$ is mapped into $(0, 1] \times \{0\}$ because of (7.183), thus (7.178) is true, and the proof is complete. □

We end this chapter with the proof of Theorem 3.6.

Proof of Theorem 3.6 Since by definition $\mathcal{A}_{m,M} \subseteq \mathcal{A}$, we have that $\mathcal{V}_{m,M} \subseteq \mathcal{V}_\mathcal{A}$. Hence, we are left with proving that the latter inclusion is strict.

We start with the case $\rho < 1$. We choose

$$\varepsilon \in \left(0, \min\left\{\frac{\rho c(c + 1)}{1 + \rho c}, \frac{M(c + 1)}{M + 1}, 1\right\}\right). \tag{7.193}$$

We observe that this choice is compatible with the assumption on ε in (7.161). We note that

$$u_1 < \min\left\{\frac{\rho c(c + 1)}{1 + \rho c}, 1\right\}, \tag{7.194}$$

thanks to (7.193). Moreover, by (7.168) and the fact that $h < \frac{1}{c}$, it holds that

$$hu + p = h(u - u_1) + hu_1 + p$$
$$= h(u - u_1) + \frac{u_1}{c} + \frac{1-\rho}{1+\rho c} \qquad (7.195)$$
$$< \frac{u}{c} + \frac{1-\rho}{1+\rho c}$$

for all $u > u_1$.

Now we choose

$$\bar{u} \in \left(u_1, \min\left\{\frac{\rho c(c+1)}{1+\rho c}, 1\right\}\right),$$

which is possible thanks to (7.194), and

$$\bar{v} := \frac{1}{2}(h\bar{u} + p) + \frac{1}{2}\left(\frac{\bar{u}}{c} + \frac{1-\rho}{1+\rho c}\right). \qquad (7.196)$$

By (7.195) we get that

$$h\bar{u} + p < \frac{1}{2}(h\bar{u} + p) + \frac{1}{2}\left(\frac{\bar{u}}{c} + \frac{1-\rho}{1+\rho c}\right) = \bar{v} < \frac{\bar{u}}{c} + \frac{1-\rho}{1+\rho c}. \qquad (7.197)$$

Using Proposition 7.5 and (7.197), we deduce that $(\bar{u}, \bar{v}) \notin \mathcal{V}_{m,M}$. By Theorem 3.3 and (7.197), we obtain instead that $(\bar{u}, \bar{v}) \in \mathcal{V}_A$. Hence, the set $\mathcal{V}_{m,M}$ is strictly included in $\mathcal{V}_A$ when $\rho < 1$.

Now we consider the case $\rho > 1$, using again the notation of Proposition 7.5. We recall that $u_2 > 0$ and $u_\infty > 0$, due to (3.13) and (7.180); hence we can choose

$$\bar{u} \in (0, \min\{u_2, u_\infty\}).$$

We also define

$$\bar{v} := \frac{1}{2}\left(\frac{1}{c} + k\right)\bar{u}.$$

By (7.186), we get that

$$k\bar{u} < \frac{k\bar{u}}{2} + \frac{\bar{u}}{2c} = \bar{v} < \frac{\bar{u}}{c}. \qquad (7.198)$$

Exploiting this and the characterization in Proposition 7.5, it holds that $(\bar{u}, \bar{v}) \notin \mathcal{V}_{m,M}$.

On the other hand, by Theorem (3.3) and (7.198) we have instead that $(\bar{u}, \bar{v}) \in \mathcal{V}_A$. As a consequence, the set $\mathcal{V}_{m,M}$ is strictly contained in $\mathcal{V}_A$ for $\rho > 1$. This concludes the proof of Theorem 3.6. □

7.6 Minimization of the War Duration

We now deal with the strategies leading to the quickest possible victory of the first population.

Proof of Theorem 3.7 Our aim is to establish the existence of the strategy leading to the quickest possible victory and to determine its range. For this, we consider the following minimization problem under constraints for $x(t) := (u(t), v(t))$:

$$\begin{cases} \dot{x}(t) = f(x(t), a(t)), \\ x(0) = (u_0, v_0), \\ x(T_s) \in (0, 1] \times \{0\}, \\ \min_{a(t) \in [m, M]} \int_0^{T_s} 1 \, dt, \end{cases} \quad (7.199)$$

where

$$f(x, a) := \Big(u(1 - u - v - ac), \; \rho v(1 - u - v) - au \Big).$$

Here T_s corresponds to the exit time introduced in (3.1), in dependence of the strategy $a(\cdot)$.

Theorem 6.15 in [92] assures the existence of a minimizing solution $(\tilde{a}, \tilde{x})$ with $\tilde{a}(t) \in [m, M]$ for all $t \in [0, T]$, and $\tilde{x}(t) \in [0, 1] \times [0, 1]$ absolutely continuous, such that $\tilde{x}(T) = (\tilde{u}(T), 0)$ with $\tilde{u}(T) \in [0, 1]$, where T is the exit time for $\tilde{a}$.

We now prove that

$$\tilde{u}(T) > 0. \quad (7.200)$$

Indeed, if this were false, then $(\tilde{u}(T), \tilde{v}(T)) = (0, 0)$. Let us call $d(t) := \tilde{u}^2(t) + \tilde{v}^2(t)$. Then, we observe that the function $d(t)$ satisfies the following differential inequality:

$$-\dot{d}(t) \leqslant Cd, \quad \text{for} \quad C := 4 + 4\rho + 2Mc + M. \quad (7.201)$$

To check this, we compute that

$$-\dot{d} = 2\Big(-\tilde{u}^2(1 - \tilde{u} - \tilde{v} - \tilde{a}c) - \tilde{v}^2 \rho(1 - \tilde{u} - \tilde{v}) + \tilde{u}\tilde{v}\tilde{a}\Big)$$
$$\leqslant 2\tilde{u}^2(2 + Mc) + 4\rho\tilde{v}^2 + (\tilde{u}^2 + \tilde{v}^2)M$$

$$\leqslant C(\tilde{u}^2 + \tilde{v}^2)$$
$$= Cd,$$

which proves (7.201).

From (7.201), one has that

$$0 < (u_0^2 + v_0^2)e^{-CT} \leqslant d(T) = \tilde{u}^2(T) + \tilde{v}^2(T) = \tilde{u}^2(T),$$

and this leads to (7.200), as desired. We remark that, in this way, we have found a trajectory $\tilde{a}$ which leads to the victory of the first population in the shortest possible time.

Theorem 6.15 in [92] assures that $\tilde{a}(t) \in L^1[0, T]$, so $\tilde{a}(t)$ is measurable. We have that the two vectorial functions F and G, defined by

$$F(u, v) := \begin{pmatrix} u(1 - u - v) \\ \rho v(1 - u - v) \end{pmatrix} \quad \text{and} \quad G(u, v) := \begin{pmatrix} -cu \\ -u \end{pmatrix},$$

and satisfying $f(x(t), a(t)) = F(x(t)) + a(t)G(x(t))$, are analytic.

Moreover the set $\overline{\mathcal{V}}_{\mathcal{A}_{m,M}}$ is a subset of $\mathbb{R}^2$; therefore it can be seen as an analytic manifold with border which is also a compact set.

For all $x_0 \in \mathcal{V}_{\mathcal{A}_{m,M}}$ and $t > 0$, we have that the trajectory starting from x_0 satisfies $x(\tau) \in \overline{\mathcal{V}}_{\mathcal{A}_{m,M}}$ for all $\tau \in [0, t]$.

Then, by Theorem 3.1 in [86], there exists a couple $(\tilde{a}, \tilde{x})$ analytic a part from a finite number of points, such that $(\tilde{a}, \tilde{x})$ solves (7.199).

Now, to study the range of $\tilde{a}$, we apply the Pontryagin Maximum Principle (see for example [92] or the original book [80]). The Hamiltonian associated with system (7.199) is

$$H(x, p, p_0, a) := p \cdot f(x, a) + p_0,$$

where $p = (p_u, p_v)$ is the adjoint to $x = (u, v)$ and p_0 is the adjoint to the cost function identically equal to 1.

The Pontryagin Maximum Principle tells us that, since $\tilde{a}(t)$ and $\tilde{x}(t) = (\tilde{u}(t), \tilde{v}(t))$ give the optimal solution, there exist a vectorial function $\tilde{p} : [0, T] \to \mathbb{R}^2$ and a scalar $\tilde{p}_0 \in (-\infty, 0]$ such that

$$\begin{cases} \dfrac{d\tilde{x}}{dt}(t) = \dfrac{\partial H}{\partial p}(\tilde{x}(t), \tilde{p}(t), \tilde{p}_0, \tilde{a}(t)), & \text{for a.a. } t \in [0, T], \\[2mm] \dfrac{d\tilde{p}}{dt}(t) = -\dfrac{\partial H}{\partial x}(\tilde{x}(t), \tilde{p}(t), \tilde{p}_0, \tilde{a}(t)), & \text{for a.a. } t \in [0, T], \end{cases} \quad (7.202)$$

7.6 Minimization of the War Duration

and

$$H(\tilde{x}(t), \tilde{p}(t), \tilde{p}_0, \tilde{a}(t)) = \max_{a(\cdot) \in [m,M]} H(\tilde{x}(t), \tilde{p}(t), \tilde{p}_0, a) \tag{7.203}$$

for a.a. $t \in [0, T]$.

Moreover, since the final time is free, we have

$$H(\tilde{x}(T), \tilde{p}(T), \tilde{p}_0, \tilde{a}(T)) = 0. \tag{7.204}$$

Also, since $H(x, p, p_0, a)$ does not depend on t, we get

$$H(\tilde{x}(t), \tilde{p}(t), \tilde{p}_0, \tilde{a}(t)) = \text{constant} = 0, \quad \text{for a.a. } t \in [0, T], \tag{7.205}$$

where the value of the constant is given by (7.204). By substituting the values of $f(x, a)$ in $H(x, p, p_0, a)$ and using (7.205), we get, for a.a. $t \in [0, T]$,

$$\tilde{p}_u \tilde{u}(1 - \tilde{u} - \tilde{v} - \tilde{a}c) + \tilde{p}_v \rho \tilde{v}(1 - \tilde{u} - \tilde{v}) - \tilde{p}_v \tilde{a} \tilde{u} + \tilde{p}_0 = 0,$$

where $\tilde{p} = (\tilde{p}_u, \tilde{p}_v)$.

Also, by (7.203) we get that

$$\max_{a \in [m,M]} H(\tilde{x}(t), \tilde{p}(t), \tilde{p}_0, a)$$
$$= \max_{a \in [m,M]} \left[-a\tilde{u}(c\tilde{p}_u + \tilde{p}_v) + \tilde{p}_u \tilde{u}(1 - \tilde{u} - \tilde{v}) + \tilde{p}_v \rho \tilde{v}(1 - \tilde{u} - \tilde{v}) + \tilde{p}_0 \right]. \tag{7.206}$$

Thus, to maximize the term in the square brackets, we must choose appropriately the value of $\tilde{a}$ depending on the sign of $\varphi(t) := c\tilde{p}_u(t) + \tilde{p}_v(t)$, that is, we choose

$$\tilde{a}(t) := \begin{cases} m & \text{if } \varphi(t) > 0, \\ M & \text{if } \varphi(t) < 0. \end{cases} \tag{7.207}$$

When $\varphi(t) = 0$, we are for the moment free to choose $\tilde{a}(t) := a_s(t)$ for every $a_s(\cdot)$ with range in $[m, M]$, without affecting the maximization problem in (7.206).

Our next goal is to determine that $a_s(t)$ has the expression stated in (3.15) for a.a. $t \in [0, T] \cap \{\varphi = 0\}$.

To this end, we claim that

$$\dot{\varphi}(t) = 0 \text{ a.e. } t \in [0, T] \cap \{\varphi = 0\}. \tag{7.208}$$

Indeed, by (7.202), we know that $\tilde{p}$ is Lipschitz continuous in $[0, T]$, hence almost everywhere differentiable, and thus the same holds for φ.

Therefore, up to a set of null measure, given $t \in [0, T] \cap \{\varphi = 0\}$, we can suppose that t is not an isolated point in such a set, and that φ is differentiable at t.

That is, there exists an infinitesimal sequence h_j for which $\varphi(t + h_j) = 0$ and

$$\dot{\varphi}(t) = \lim_{j \to +\infty} \frac{\varphi(t + h_j) - \varphi(t)}{h_j} = \lim_{j \to +\infty} \frac{0 - 0}{h_j} = 0,$$

and this establishes (7.208).

Consequently, in light of (7.208), a.a. $t \in [0, T] \cap \{\varphi = 0\}$ satisfies

$$0 = \dot{\varphi}(t)$$
$$= c\frac{d\tilde{p}_u}{dt}(t) + \frac{d\tilde{p}_v}{dt}(t)$$
$$= c\big[-\tilde{p}_u(t)(1 - 2\tilde{u}(t) - \tilde{v}(t) - ca_s(t)) + \tilde{p}_v(t)(\rho\tilde{v}(t) + a_s(t))\big]$$
$$+ \tilde{p}_u(t)\tilde{u}(t) - \tilde{p}_v(t)\rho(1 - \tilde{u}(t) - 2\tilde{v}(t)).$$

Now, since $\varphi(t) = 0$, we have that $\tilde{p}_v(t) = -c\tilde{p}_u(t)$; inserting this information in the last equation, we get

$$\begin{aligned} 0 = {} & -\tilde{p}_u c(1 - 2\tilde{u} - \tilde{v} - a_s c) - \tilde{p}_u \rho c^2 \tilde{v} \\ & - \tilde{p}_u a_s c^2 + \tilde{p}_u \tilde{u} + \tilde{p}_u \rho(1 - \tilde{u} - 2\tilde{v}). \end{aligned} \quad (7.209)$$

Notice that if $\tilde{p}_u = 0$, then $\tilde{p}_v = -c\tilde{p}_u = 0$; moreover, by (7.205), one gets $\tilde{p}_0 = 0$. But by the Pontryagin Maximum Principle, one cannot have $(\tilde{p}_u, \tilde{p}_v, \tilde{p}_0) = (0, 0, 0)$, and therefore one obtains $\tilde{p}_u \neq 0$ in $\{\varphi = 0\}$.

Hence, dividing (7.209) by $\tilde{p}_u$ and rearranging the terms, one gets

$$\tilde{u}(2c + 1 - \rho c) + c\tilde{v}(1 - \rho c - 2\rho) + c(\rho - 1) = 0. \quad (7.210)$$

Differentiating the expression in (7.210) with respect to time, we get

$$\tilde{u}(2c + 1 - \rho c)(1 - \tilde{u} - \tilde{v} - ac) + c(1 - \rho c - 2\rho)[\rho\tilde{v}(1 - \tilde{u} - \tilde{v}) - a\tilde{u}] = 0$$

that yields

$$a_s = \frac{(1 - \tilde{u} - \tilde{v})(\tilde{u}(2c + 1 - \rho c) + \rho c)}{2c\tilde{u}(c + 1)},$$

which is the desired expression. By a slight abuse of notation, we define the function $a_s(t) = a_s(\tilde{u}(t), \tilde{v}(t))$ for $t \in [0, T]$. Notice that since $\tilde{u}(t) > 0$ for $t \in [0, T]$, $a_s(t)$ is continuous for $t \in [0, T]$. $\square$

Bibliographical Notes

We give here some references to the classical books, where the reader can find more detailed explanations of the theories we exploit, and some guidance in the literature of Lotka–Volterra systems.

We start with some classical references to dynamical systems. In this book, we often refer to the book of Perko [77], which covers those topics necessary for a clear understanding of the qualitative theory of ordinary differential equations and the concept of a dynamical system. It is written for advanced undergraduates and for beginning graduate students, and it is useful as a primer for young people interested in research in dynamical systems. In particular, it focuses on describing the qualitative behavior of the solution set of a given system of differential equations, including the invariant sets and limiting behavior of the dynamical system. This includes the Stable Manifold Theorem and the Pointcaré–Bendixon Theorem, which we often exploit.

Another excellent reference is the book [46] by Hirsh and Smale. The advantage of this book is the vast collection of examples, including the predatory–prey and the SIR model, which are presented to visualize the theorems and their hypothesis. This text is more accessible to non-mathematicians and undergraduate students, being at the same time very complete and satisfactory as a background for our text.

The book [100] by Wiggings takes a step forward. In the first chapters, it presents the geometrical analysis of dynamical systems already covered in [46], but in a more general and abstract form. Then, it continues with advanced material, working with Hamiltonians and their relations with the other tools, and exposing questions that are relevant in research. The text is intended for an audience with "mathematical maturity" and interested in research on dynamical systems. The author says that the material is abundant and ambitious even for a three terms program.

Pontryagin's Maximum Principle was first presented in the book [80] by Pontryagin, Boltayanskii, Gamkrelidze, and Mishchenko, in the form of multiple theorems based on the same principle. This text is however hard to read, and nowadays more accessible texts are available.

For control theory and optimization, the textbook [92] by Trélat is a very good starting point. Targeted at undergraduate students, it provides a clear and concise introduction to problems and tools in control theory, including controllability and Pontryagin Maximum Principle, and it is completed by examples and numerical methods. The proposed material is enough background for the comprehension of our book. We can also recommend [67] by Macki and Strauss as an advanced undergraduate text. Other reliable options are the lecture notes [31] by Férnandez Cara and Zuazua or [69] by Micu and Zuazua.

More advanced books in control theory focus on geometric control, a branch that has been developed since the 1960s using the tools of differential geometry and Lie theory to prescribe optimal strategies, called optimal feedback, for a dynamic optimization problem. To the scope of our book, a great reference is the book [16] by Boscain and Piccoli. In fact, their work focuses on time minimization problems in 2-D systems, for which the authors provide a complete theory, based on the research papers of the early 2000s.

For a comprehensive treatment of geometric control applied to both linear and nonlinear systems, including stabilization, consider the references [23] by Coron and [5] Agrachev and Sachkov.

Control theory also originated the branch of game theory that analyzes systems in which one or more parties take strategic decisions in order to achieve an optimal situation for their interests. Started with a finite number of players taking decisions from a discrete set of choices, this fascinating field has evolved in various ways. When the situation evolves according to a dynamical system, where some of the parameters represent the strategic decisions of the players, we talk about differential games. An overview of the connection between control theory and differential games can be found for instance in the books [7, 28], see also [29] where the subject is approached within the theory of viscosity solutions. We also refer to the book [53], which focuses in particular on some differential games modeling war conflicts.

The classic Lotka–Volterra equations for modeling predator–prey systems were first introduced independently in [65] and [95] in the 2020s. The work of Volterra [95] focused on three behaviors—the predation, the competition for the same resources, and the mutualism. In the same decades, the emergence of numerous models inspired by ecology gave birth to the branch of mathematics that we call mathematical biology. In the 1980s, Murray wrote the milestone books [71, 72] to bring together the fundamentals of this new subject. We also cite the text [17] by Cantarell and Cosner, which collects the rigorous mathematics formalizing many of the insights that have been had throughout the twentieth century in mathematical biology. However, this book treats mainly single-species equations. In fact, ecological modelization has grown so much that it is impossible to give a complete picture.

A series of six papers by Hirsh and Morris answered many open questions on the dynamics of competitive and collaborative systems. The ones regarding two-dimensional models are [43–45].

Bibliographical Notes

On the other hand, the Lotka–Volterra model was extended in different ways. A very successful additive feature is the spatial diffusion of the individuals of the two species. As examples of studies on the subject, a long list of works on traveling waves for Lotka–Volterra competition systems are available, see [35] by Gardner, [88] by Tang and Fife, [55] by Kan-On, [39] by Guo and Lin just to cite a few. Spatial segregation caused by competition was studied by Tavares, Terracini, Verzini and their collaborators in [22, 89] and the following works.

To conclude, from the viewpoint of our book, another relevant use of Lotka–Volterra competitive systems consists of applications to diffusion of new technologies that substitute old ones. The Bass model introduced in [8] became one of the most influential in Management Science of the last century, see also [59] and the subsequent research papers.

References

1. Y. Achdou, F.J. Buera, J.-M. Lasry, P.-L. Lions, B. Moll, Partial differential equation models in macroeconomics. Philos. Trans. R. Soc. Lond. Ser. A Math. Phys. Eng. Sci. **372**(2028), 20130397 (2014)
2. R.E.W. Adams, *Prehistoric Mesoamerica* (University of Oklahoma Press, Norman, 1991)
3. T. Alfred, *Wasáse: Indigenous Pathways of Action and Freedom* (University of Toronto Press, Toronto, 2005)
4. T. andrea, Kinetic equations and stochastic game theory for social systems, in *Mathematical Models and Methods for Planet Earth*. Springer INdAM Series, vol. 6 (Springer, Cham, 2014), pp. 37–57
5. A.A. Agrachev, Y. Sachkov, *Control Theory from the Geometric Viewpoint*, vol. 87 (Springer Science & Business Media, Berlin, 2013)
6. J.S. Bain, *Barriers to New Competition: Their Character and Consequences in Manufacturing Industries* (Harvard University Press, Cambridge, 1956)
7. M. Bardi, I. Capuzzo-Dolcetta, *Optimal Control and Viscosity Solutions of Hamilton-Jacobi-Bellman Equations*. (Birkhäuser Boston Inc., Boston, 1997). With appendices by Maurizio Falcone and Pierpaolo Soravia
8. F.M. Bass, A new product growth for model consumer durables. Manage. Sci. **15**(5), 215–227 (1969)
9. A.D. Bazykin, Nonlinear dynamics of interacting populations, in *World Scientific Series on Nonlinear Science. Series A: Monographs and Treatises*, vol. 11 (World Scientific Publishing Co., Inc., River Edge, 1998). With a biography of the author by E.P. Kryukova, Y.A. Bazykin, D.A. Bazykin, Edited and with a foreword by A.I. Khibnik, B. Krauskopf
10. N. Bellomo, F. Brezzi, Challenges in active particles methods: theory and applications. Math. Models Methods Appl. Sci. **28**(9), 1627–1633 (2018)
11. N. Bellomo, P. Degond, E. Tadmor (eds.), *Active Particles*, vol. 2. *Advances in Theory, Models and Applications*. Modeling and Simulation in Science, Engineering and Technology (Birkhäuser/Springer, Cham, 2019)
12. H. Berestycki, S. Nordmann, L. Rossi, Modeling the propagation of riots, collective behaviors and epidemics. Math. Eng. **4**(1), Paper No. 003, 53 (2022)
13. H. Berestycki, J. Wei, M. Winter, Existence of symmetric and asymmetric spikes for a crime hotspot model. SIAM J. Math. Anal. **46**(1), 691–719 (2014)
14. S.C. Bhargava, Generalized Lotka–Volterra equations and the mechanism of technological substitution. Techn. Forecasting Soc. Change **35**(4), 319–326 (1989)
15. D. Bonaldo, *Competizione Tra Prodotti Farmaceutici: Strumenti di Previsione*. PhD thesis, Master's thesis (University of Padua, Padua, 1991)

16. U. Boscain, B. Piccoli, *Optimal Syntheses for Control Systems on 2-D Manifolds*, vol. 43 (Springer Science & Business Media, New York, 2003)
17. R.S. Cantrell, C. Cosner, *Spatial Ecology via Reaction-Diffusion Equations* (Wiley, New York, 2004)
18. J. Carr, Applications of centre manifold theory, in *Applied Mathematical Sciences*, vol. 35 (Springer, New York, 1981)
19. A.S. Chakrabarti, Stochastic Lotka-Volterra equations: a model of lagged diffusion of technology in an interconnected world. Phys. A **442**, 214–223 (2016)
20. P. Collier, A. Hoeffler, Greed and grievance in civil war, in *World Bank Policy Res. Working Paper, 2355* (2002)
21. P. Collier, N. Sambanis, *Understanding Civil War, Evidence and Analysis. Africa*, vol. 1 (World Bank Publications, New York, 2005)
22. M. Conti, S. Terracini, G. Verzini, Asymptotic estimates for the spatial segregation of competitive systems. Adv. Math. **195**(2), 524–560 (2005)
23. J.-M. Coron, Control and nonlinearity, in *Mathematical Surveys and Monographs*, vol. 136 (American Mathematical Society, Providence, RI, 2007)
24. J. Cramer, The early origins of the logit model. Studies in Hist. Phil. Sci. Part C: Biol. Biomed. Sci. **35**(4), 613–626 (2004)
25. E.C.M. Crooks, E.N. Dancer, D. Hilhorst, M. Mimura, H. Ninomiya, Spatial segregation limit of a competition-diffusion system with Dirichlet boundary conditions. Nonlinear Anal. Real World Appl. **5**(4), 645–665 (2004)
26. N.B. Davies, J.R. Krebs, S.A. West, *An Introduction to Behavioural Ecology* (Wiley, New York, 2012)
27. R. Dawkins, *The Selfish Gene* (Oxford University Press, Oxford, 2016)
28. E.J. Dockner, S. Jorgensen, N.V. Long, G. Sorger, et al. Differential games in economics and management science, in *Cambridge Books* (2000)
29. L.C. Evans, P.E. Souganidis, Differential games and representation formulas for solutions of Hamilton-Jacobi-Isaacs equations. Indiana Univ. Math. J. **33**(5), 773–797 (1984)
30. B.D. Fath, *Encyclopedia of Ecology* (Elsevier, Amsterdam, 2018)
31. E. Fernández Cara, E. Zuazua Iriondo, Control theory: history, mathematical achievements and perspectives. Boletín de la Sociedad Española de Matemática Aplicada **26**, 79–140 (2003)
32. K. Fischbach, J. Marx, T. Weitzel, Agent-based modeling in social sciences. J. Busin. Econ. **91**, 1263–1270 (2021)
33. J. Flores, A mathematical model for Neanderthal extinction. J. Theoret. Biol. **191**(3), 295–298 (1998)
34. C.A. Floudas, P.M. Pardalos (eds.), *Optimization in Computational Chemistry and Molecular Biology*. Nonconvex Optimization and its Applications, vol. 40 (Kluwer Academic Publishers, Dordrecht, 2000). Local and global approaches, Papers from the conference held at Princeton University, Princeton, NJ, May 7–9, 1999
35. R.A. Gardner, Existence and stability of travelling wave solutions of competition models: a degree theoretic approach. J. Differ. Equ. **44**(3), 343–364 (1982)
36. S. Gaucel, M. Langlais, D. Pontier, Invading introduced species in insular heterogeneous environments. Ecol. Modell. **188**(1), 62–75 (2005)
37. J.R. Graef, J. Henderson, L. Kong, X.S. Liu, *Ordinary Differential Equations and Boundary Value Problems*, vol. I. Trends in Abstract and Applied Analysis, vol. 7 (World Scientific Publishing Co. Pte. Ltd., Hackensack, NJ, 2018). Advanced ordinary differential equations
38. R.E. Green, J. Krause, T. Maricic, U. Stenzel, M. Kircher, N. Patterson, H. Li, W. Zhai, A draft sequence of the Neanderthal genome. Science **328**, 710–722 (2010)

39. J.-S. Guo, Y.-C. Lin, The sign of the wave speed for the Lotka-Volterra competition-diffusion system. Commun. Pure Appl. Anal **12**(5), 2083–2090 (2013)
40. O.P. Heil, K. Helsen, Toward an understanding of price wars: their nature and how they erupt. Int. J. Res. Mark. **18**(1), 83–98 (2001). Competition and Marketing
41. T. Higham, K. Douka, R. Wood, C.B. Ramsey, F. Brock, L. Basell, M. Camps, A. Arrizabalaga, J. Baena, C. Barroso-Ruíz, C. Bergman, C. Boitard, P. Boscato, M. Caparrós, N.J. Conard, C. Draily, A. Froment, B. Galván, P. Gambassini, A. Garcia-Moreno, S. Grimaldi, P. Haesaerts, B. Holt, M.-J. Iriarte-Chiapusso, A. Jelinek, J.F. Jordá Pardo, J.-M. Maíllo-Fernández, A. Marom, J. Maroto, M. Menéndez, L. Metz, E. Morin, A. Moroni, F. Negrino, E. Panagopoulou, M. Peresani, S. Pirson, M. de la Rasilla, J. Riel-Salvatore, A. Ronchitelli, D. Santamaria, P. Semal, L. Slimak, J. Soler, N. Soler, A. Villaluenga, R. Pinhasi, R. Jacobi, The timing and spatiotemporal patterning of Neanderthal disappearance. Nature **512**, 306–309 (2014)
42. A. Hironaka, *Neverending Wars: The International Community, Weak States and the Perpetuation of Civil War* (Harvard University Press, Cambridge, 2005)
43. M.W. Hirsch, Systems of differential equations which are competitive or cooperative: I. limit sets. SIAM J. Math. Anal. **13**(2), 167–179 (1982)
44. M.W. Hirsch, Systems of differential equations that are competitive or cooperative II: Convergence almost everywhere. SIAM J. Math. Anal. **16**(3), 423–439 (1985)
45. M.W. Hirsch, Systems of differential equations which are competitive or cooperative: III. Competing species. Nonlinearity **1**(1), 51 (1988)
46. M.W. Hirsch, S. Smale, Differential equations, dynamical systems and linear algebra, in *Pure and Applied Mathematics*, vol. 60 (Academic Press [Harcourt Brace Jovanovich, Publishers], New York, 1974)
47. B. Hölldobler, E.O. Wilson, et al. *The Ants* (Harvard University Press, Cambridge, 1990)
48. W. Horsthemke, Noise induced transitions, in *Nonequilibrium Dynamics in Chemical Systems (Bordeaux, 1984)*. Springer Series Synergetics, vol. 27 (Springer, Berlin, 1984), pp. 150–160
49. S.-B. Hsu, *Ordinary Differential Equations with Applications*. Series on Applied Mathematics, vol. 21, 2nd edn. (World Scientific Publishing Co. Pte. Ltd., Hackensack, NJ, 2013).
50. M. Huang, R.P. Malhamé, P.E. Caines, Large population stochastic dynamic games: closed-loop McKean-Vlasov systems and the Nash certainty equivalence principle. Commun. Inf. Syst. **6**(3), 221–251 (2006)
51. J.-J. Hublin, N. Sirakov, V. Aldeias, S. Bailey, E. Bard, V. Delvigne, E. Endarova, Y. Fagault, H. Fewlass, M. Hajdinjak, B. Kromer, I. Krumov, J.a. Marreiros, N.L. Martisius, L. Paskulin, V. Sinet-Mathiot, M. Meyer, S. Pääbo, V. Popov, Z. Rezek, S. Sirakova, M.M. Skinner, G.M. Smith, R. Spasov, S. Talamo, T. Tuna, L. Wacker, F. Welker, A. Wilcke, N. Zahariev, S.P. McPherron, T. Tsanova, Initial upper Palaeolithic Homo sapiens from Bacho Kiro Cave, Bulgaria. Nature **581**, 299–302 (2020)
52. P.A. Iglesias, B.P. Ingalls (eds.), *Control Theory and Systems Biology* (MIT Press, Cambridge, MA, 2010)
53. R. Isaacs, *Differential Games: A Mathematical Theory with Applications to Warfare and Pursuit, Control and Optimization* (Courier Corporation, North Chelmsford, 1999)
54. S. Kalish, A new product adoption model with price, advertising and uncertainty. Manage. Sci. **31**(12), 1569–1585 (1985)
55. Y. Kan-On, Fisher wave fronts for the Lotka-Volterra competition model with diffusion. Nonlinear Anal. Theory Methods Appl. **28**(1), 145–164 (1997)
56. L.H. Keeley, *War Before Civilization* (Oxford University Press, Oxford, 1996)
57. D. Keen, Incentives and disincentives for violence, in *Greed and Grievance: Economic Agendas in Civil Wars*, ed. by M. Berdal, D. Malone (Boulder, Lynne Rienner, 2000), pp. 19–43

58. W.O. Kermack, A.G. McKendrick, A contribution to the mathematical theory of epidemics. Proc. R. Soc. Lond. A **115**, 700–721 (1927)
59. T.V. Krishnan, F.M. Bass, V. Kumar, Impact of a late entrant on the diffusion of a new product/service. J. Mark. Res. **37**(2), 269–278 (2000)
60. S. Kuhn, M. Stiner, What's a mother to do? The division of labor among Neanderthals and modern humans in Eurasia. Current Anthrop. **47**(6), 953–981 (2006)
61. L.R. Kurtz, *Encyclopedia of Violence, Peace and Conflict, Three-Volume Set* (Academic Press, New York, 1999)
62. J.-M. Lasry, P.-L. Lions, Mean field games. Jpn. J. Math. **2**(1), 229–260 (2007)
63. S.A. LeBlanc, K.E. Register, *Constant Battles. Why We Fight* (St. Martins Press, New York, 2004)
64. K. Lorenz, *On Aggression* (Routledge, England, 1967)
65. A.J. Lotka, Analytical note on certain rhythmic relations in organic systems. Proc. Natl. Acad. Sci. USA **6**(7), 410–415 (1920)
66. A.J. Lotka, Elements of physical biology. Sci. Prog. Twent. Century (1919–1933) **21**(82), 341–343 (1926)
67. J. Macki, A. Strauss, *Introduction to Optimal Control Theory* (Springer Science & Business Media, Berlin, 2012)
68. A. Massaccesi, E. Valdinoci, Is a nonlocal diffusion strategy convenient for biological populations in competition? J. Math. Biol. **74**(1–2), 113–147 (2017)
69. S. Micu, E. Zuazua, An introduction to the controllability of partial differential equations, in *Quelques Questions de théorie du Contrôle*, ed. by T. Sari. Collection Travaux en Cours Hermann, to appear (2004)
70. S.A. Morris, D. Pratt, Analysis of the Lotka–Volterra competition equations as a technological substitution model. Techn. Forecasting Soc. Change **70**(2), 103–133 (2003)
71. J.D. Murray, *Mathematical Biology. I.* Interdisciplinary Applied Mathematics, vol. 17, 3rd edn. (Springer, New York, 2002). An introduction
72. J.D. Murray, *Mathematical Biology. II* (2003). Spatial models and biomedical applications
73. W. Nakahashi, The effect of trauma on Neanderthal culture: a mathematical analysis. HOMO **68**(2), 83–100 (2017)
74. T. Namba, M. Mimura, Spatial distribution of competing populations. J. Theoret. Biol. **87**(4), 795–814 (1980)
75. R.L. O'Connell, *Of Arms and Men: A History of War, Weapons, Aggression* (Oxford University, Oxford, 1989)
76. S. Ontañón, G. Synnaeve, A. Uriarte, F. Richoux, D. Churchill, M. Preuss, A survey of Real-Time Strategy Game AI research and competition in StarCraft. IEEE Trans. Comput. Intell. AI Games **5**(4), 293–311 (2013)
77. L. Perko, *Differential Equations and Dynamical Systems*, vol. 7 (Springer Science & Business Media, New York, 2013)
78. B. Perthame, *Transport Equations in Biology*. Frontiers in Mathematics (Birkhäuser, Basel, 2007)
79. B. Perthame, *Parabolic Equations in Biology*. Lecture Notes on Mathematical Modelling in the Life Sciences (Springer, Cham, 2015). Growth, reaction, movement and diffusion
80. L.S. Pontryagin, V.G. Boltayanskii, R.V. Gamkrelidze, E.F. Mishchenko, *Mathematical Theory of Optimal Processes* (Routledge, England, 2018)
81. M. Porfiri, G. Ariel, On effective temperature in network models of collective behavior. Chaos **26**(4), 043109 (2016)
82. K. Raaflaub, N. Rosenstein, *War and Society in the Ancient and Medieval Worlds. Asia, The Mediterranean, Europe and Mesoamerica* (Harvard University Press, Cambridge, 1999)

83. S.N. Rasband, *Chaotic Dynamics of Nonlinear Systems*. A Wiley-Interscience Publication (Wiley, New York, 1990)
84. L.F. Richardson, *Arms and Insecurity: A Mathematical Study of the Causes and Origins of War*, ed. by N. Rashevsky, E. Trucco (The Boxwood Press, Pittsburgh, PA; Quadrangle Books, Chicago, Ill, 1960)
85. T. Saito, K. Shigemoto, A logistic curve in the sir model and its application to deaths by Covid-19 in Japan. *medRxiv* (2020)
86. H.J. Sussmann, Regular synthesis for time-optimal control of single-input real analytic systems in the plane. SIAM J. Control Optim. **25**(5), 1145–1162 (1987)
87. A. Swanson, T. Arnold, M. Kosmala, J. Forester, C. Packer, In the absence of a "landscape of fear": How lions, hyenas and cheetahs coexist. Ecol. Evol. **6**(23), 8534–8545 (2016)
88. M.M. Tang, P.C. Fife, Propagating fronts for competing species equations with diffusion. Arch. Ration. Mech. Anal. **73**(1), 69–77 (1980)
89. H. Tavares, S. Terracini, Regularity of the nodal set of segregated critical configurations under a weak reflection law. Calc. Var. Partial Differ. Equations **45**, 273–317 (2012)
90. G. Teschl, *Ordinary Differential Equations and Dynamical Systems*. Graduate Studies in Mathematics, vol. 140 (American Mathematical Society, Providence, RI, 2012)
91. M.D. Toft, The state of the field: demography and war. Population and Conflict: Exploring the Links **1**(11), 25–28 (2005)
92. E. Trélat, *Contrôle Optimal: Théorie & Applications* (Vuibert, Paris, 2005)
93. J.M.G. van der Dennen, *The Origin of War: The Evolution of a Male-Coalitional Reproductive Strategy* (Origin Press, Chicago, 1995)
94. G. Vandenbroucke, Fertility and wars: the case of world war I in France. Am. Econ. J. Macroecon. **6**(2), 108–136 (2014)
95. V. Volterra, Variazioni e fluttuazioni del numero d'individui in specie animali conviventi, Mem. R. Accad. Linei Ser **6**, 31–113 (1926)
96. V. Volterra, Principes de biologie mathématique. Acta Biotheor. **3**(1), 1–36 (1937)
97. B.F. Walter, Why bad governance leads to repeat civil war. J. Conflict Resolution **59**(7), 1242–1272 (2015)
98. C. Watanabe, R. Kondo, A. Nagamatsu, Policy options for the diffusion orbit of competitive innovations—an application of Lotka–Volterra equations to Japan's transition from analog to digital TV broadcasting. Technovation **23**(5), 437–445 (2003)
99. C. Watanabe, R. Kondo, N. Ouchi, H. Wei, A substitution orbit model of competitive innovations. Techn. Forecasting Soc. Change **71**(4), 365–390 (2004)
100. S. Wiggins, *Introduction to Applied Nonlinear Dynamical Systems and Chaos*. Texts in Applied Mathematics, vol. 2 (Springer, New York, 1990)

SPRINGER NATURE

GPSR Compliance

The European Union's (EU) General Product Safety Regulation (GPSR) is a set of rules that requires consumer products to be safe and our obligations to ensure this.

If you have any concerns about our products, you can contact us on ProductSafety@springernature.com

In case Publisher is established outside the EU, the EU authorized representative is:

Springer Nature Customer Service Center GmbH
Europaplatz 3
69115 Heidelberg, Germany

The manufacturer's authorised representative in the EU is Springer Nature Customer Service Centre GmbH, Europaplatz 3, 69115 Heidelberg, Germany. If you have any concerns regarding our products, please contact ProductSafety@springernature.com

Printed and bound by CPI Group (UK) Ltd, Croydon, CR0 4YY

26/03/2026

02078980-0003